Alpaca Perspectives

Inspiration, Images and Insights from the Paddock

Amee Dennis

Daisee Press

1st edition 2025 by Daisee Press

Cover Design by Amee Dennis

Cover Photography by Nicole Drew

Internal Photography by Nicole Drew

Disclaimer

Contents

For Mel.

My oldest friend. Thank you for always looking past my broken fence to see the flowers in my garden... Even after the goat got in and ate them all.

For Miss Nicole.

A few years ago now, you took my vision and passion and brought it to life. Since then, you've nailed the perfect shot in every single photo shoot, capturing the light and essence of the farm, our alpacas and our products, no matter how outrageous the brief or impossible the timeframe.

You, are a freaking legend.

Introduction

Welcome to the farm

I didn't grow up dreaming of writing a book about alpacas.

In fact, if you'd told past-me that one day I'd be standing in a paddock, covered in dust, and questionable bodily fluids, trying to put words around the chaos of farm life — I would have laughed. Or cried. Or both.

Somewhere between the feed runs, the fences that insist on breaking themselves, the endless to-do list, and the alpacas who wake up each morning determined to test my patience, I started scribbling down little moments. Not because I thought they were important, but because I didn't want to forget them.

What I didn't realise at the time was that these small notes — the things that made me smile, swear, stop, or breathe — were quietly becoming the story of my life on the land.

People often imagine farming as peaceful or romantic. And sometimes it is. But most days it's messy, unpredictable, character building in the worst and best ways, and absolutely not for the fainthearted. The alpacas and I have survived drought, storms, heartbreak, triumphs, questionable decisions, and at least one incident that definitely involved running with one boot on.

But in all of that, there's been so much unexpected beauty.

Not the big kind — the small kind.

The kind you only notice when you slow down long enough to see it.

This book is made up of those moments.

The ridiculous ones, the tender ones, the ones that broke me open and put me back together.

The ones that taught me to keep going, pay attention, laugh at myself, and look for the small good thing in any day, no matter how chaotic.

It's part stunning photography of the main characters, part reflection, part "what on earth just happened?"

You don't need to be a farmer, or even a fan of alpacas (though it helps), to find yourself in these pages. You just need to have lived a life where things haven't gone to plan — which is most of us, really.

So wherever you are, however you found yourself holding this book, I'm glad you're here.

Turn the page.

Let the farm meet you where you are.

There's a good chance you'll find something in here you didn't know you needed. It's good like that.

Part 1

Inspiration and Images

Daisee Press

Beginnings

Notes from the Paddock

Day one of alpaca ownership.

And when I say "day one," I mean *I knew absolutely nothing.* Not a thing. They could simply have been long necked sheep for all I knew. I made my morning coffee, stood at the kitchen window, and took in the early morning light on the farm like every other normal day.

That's when I saw them.

Nine alpacas.

All lying completely flat along the driveway.

Dead. Every single one of them. Or so I thought.

The cup was thrown, the screaming started, and I bolted out the door — half in my boots, half in my pyjamas, looking like an unhinged scarecrow with morning hair. I was crying, sprinting, trying to get my arms into a jacket sleeve that absolutely did not want to cooperate.

I hit the gate at full speed, swung it open with a crash, took another two steps and froze.

All nine alpacas sat up.

Just... sat up.

And stared at me.

I stood there, panting, wild-eyed, nose running from the cold, completely undone, while they blinked at me like *"...you good?"*

That was my introduction to what I now know is perfectly normal alpaca behaviour. I still ask Granny Mae, one of those original nine, what on earth she thought of me in that moment. She still refuses to tell me, but her expression said plenty.

I didn't know it then, but that ridiculous moment was the real beginning.

Because beginnings rarely look like you think they will.

They look like chaos. And confusion. And you, standing in a paddock at dawn, learning something for the first time — usually the hard way.

Maybe that's the secret.

You don't have to know what you're doing.

You just have to open the gate.

Everything begins the moment you believe in yourself.

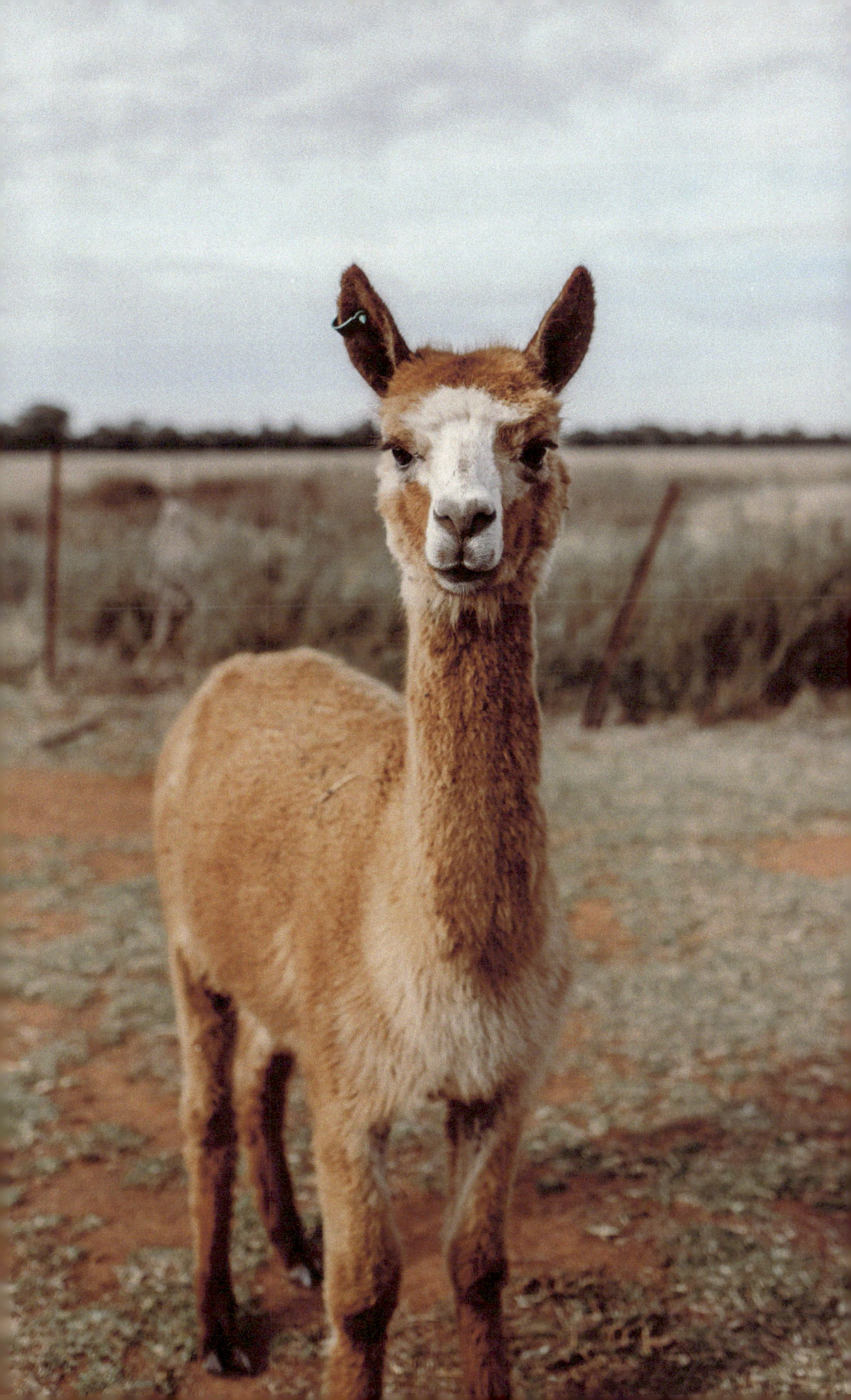

Don't wait for the perfect moment. Create one.

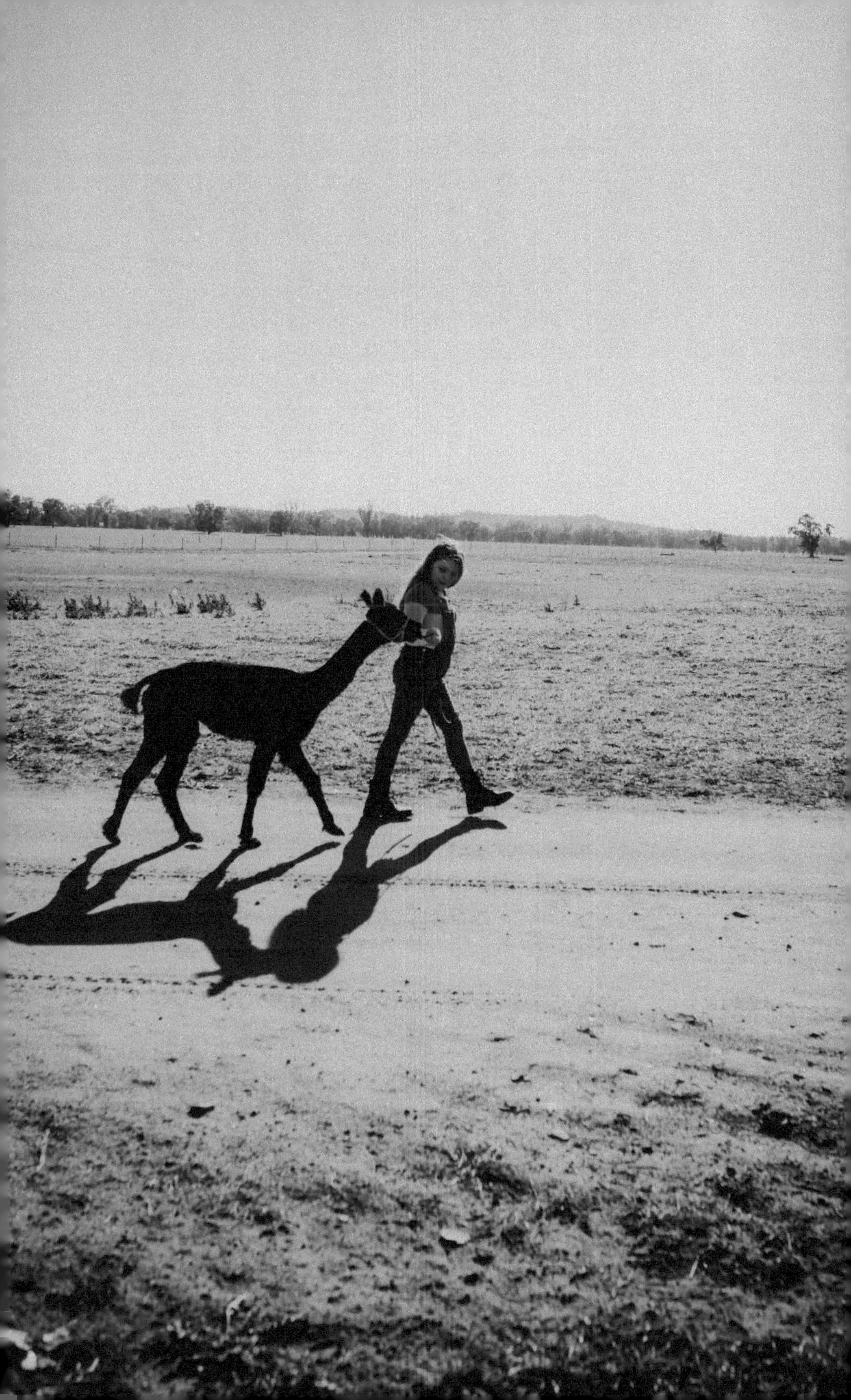

With every challenge comes a chance to bloom.

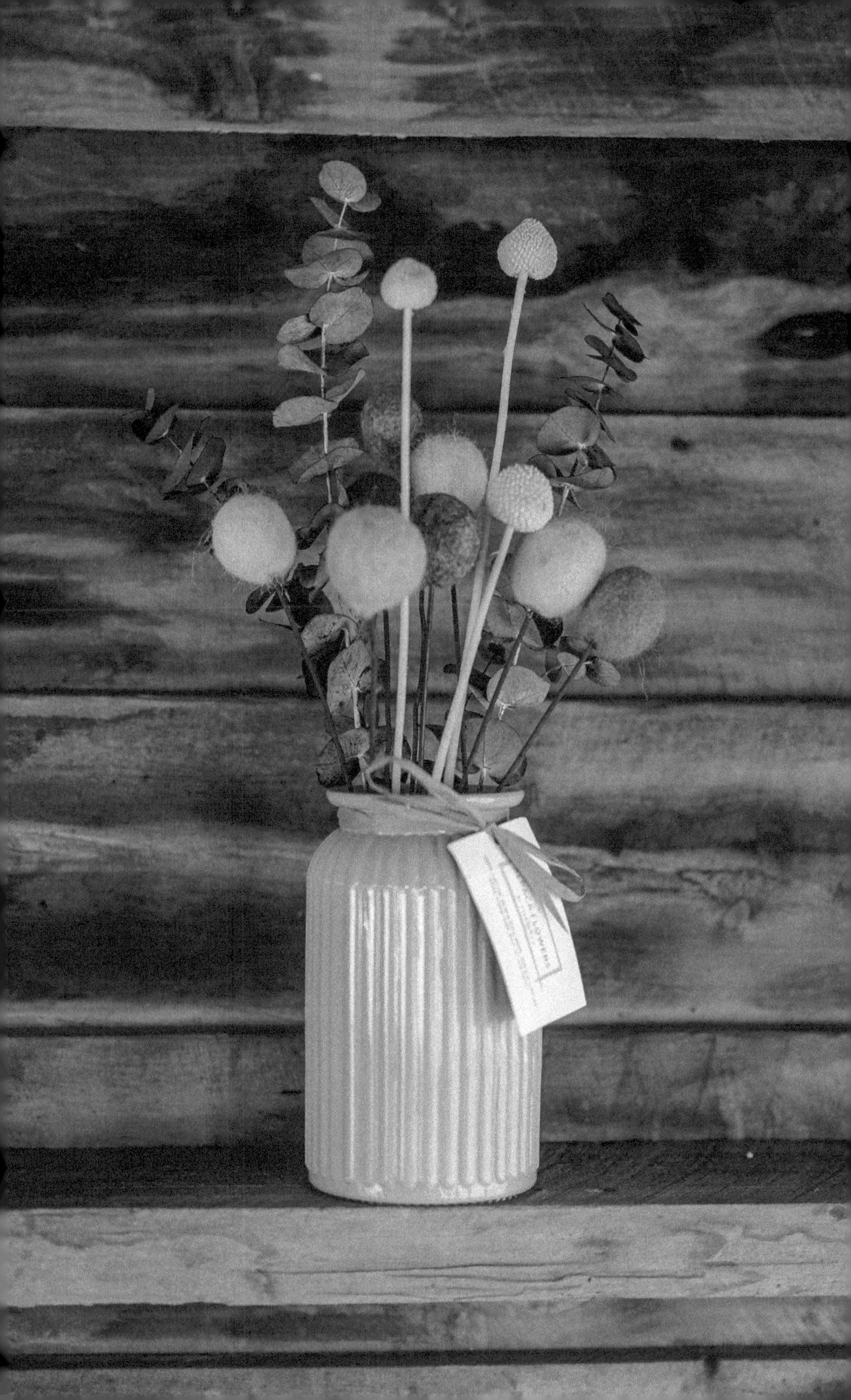

Every step forward, no matter how small, counts.

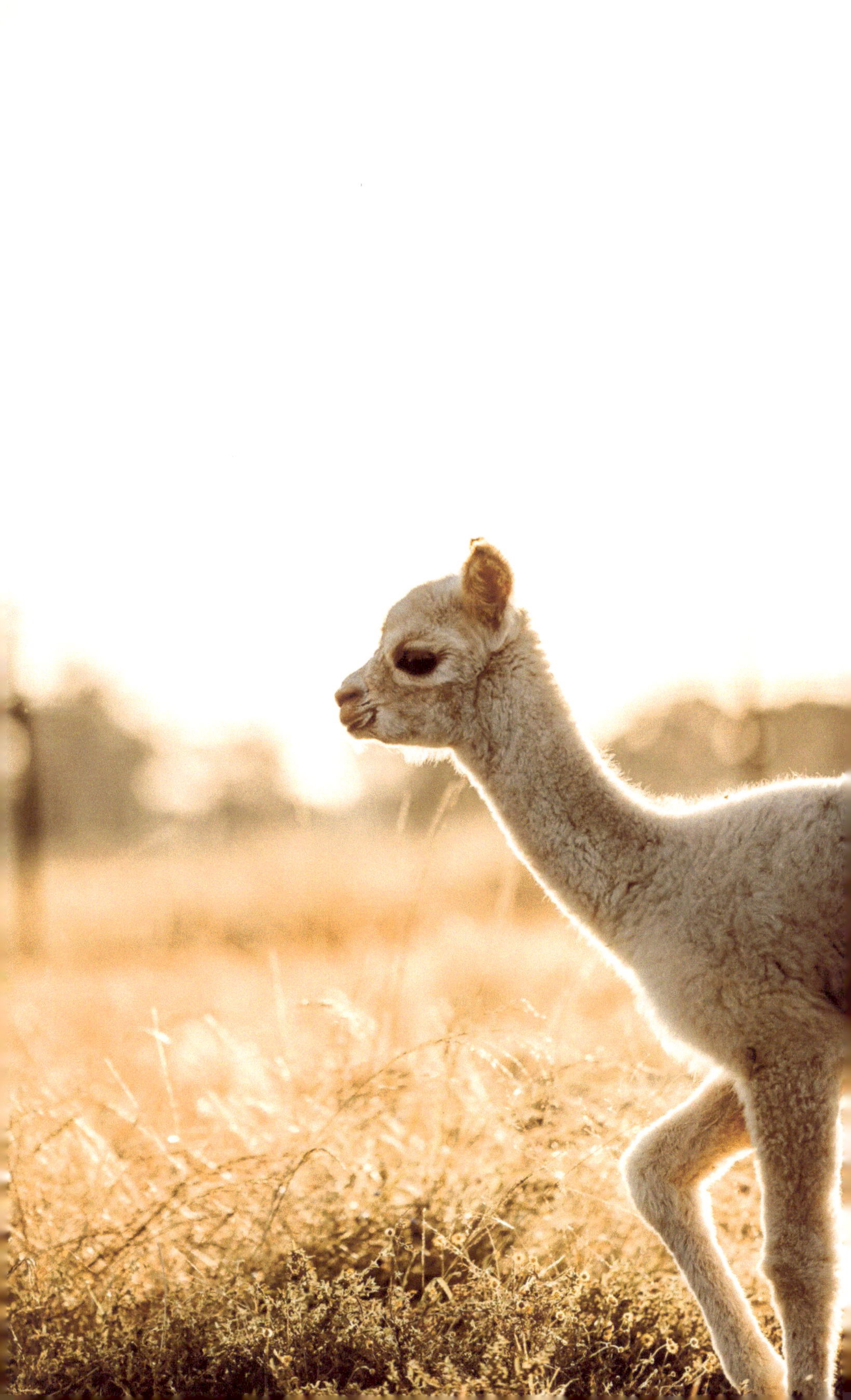

When you embrace your uniqueness, you light up the world.

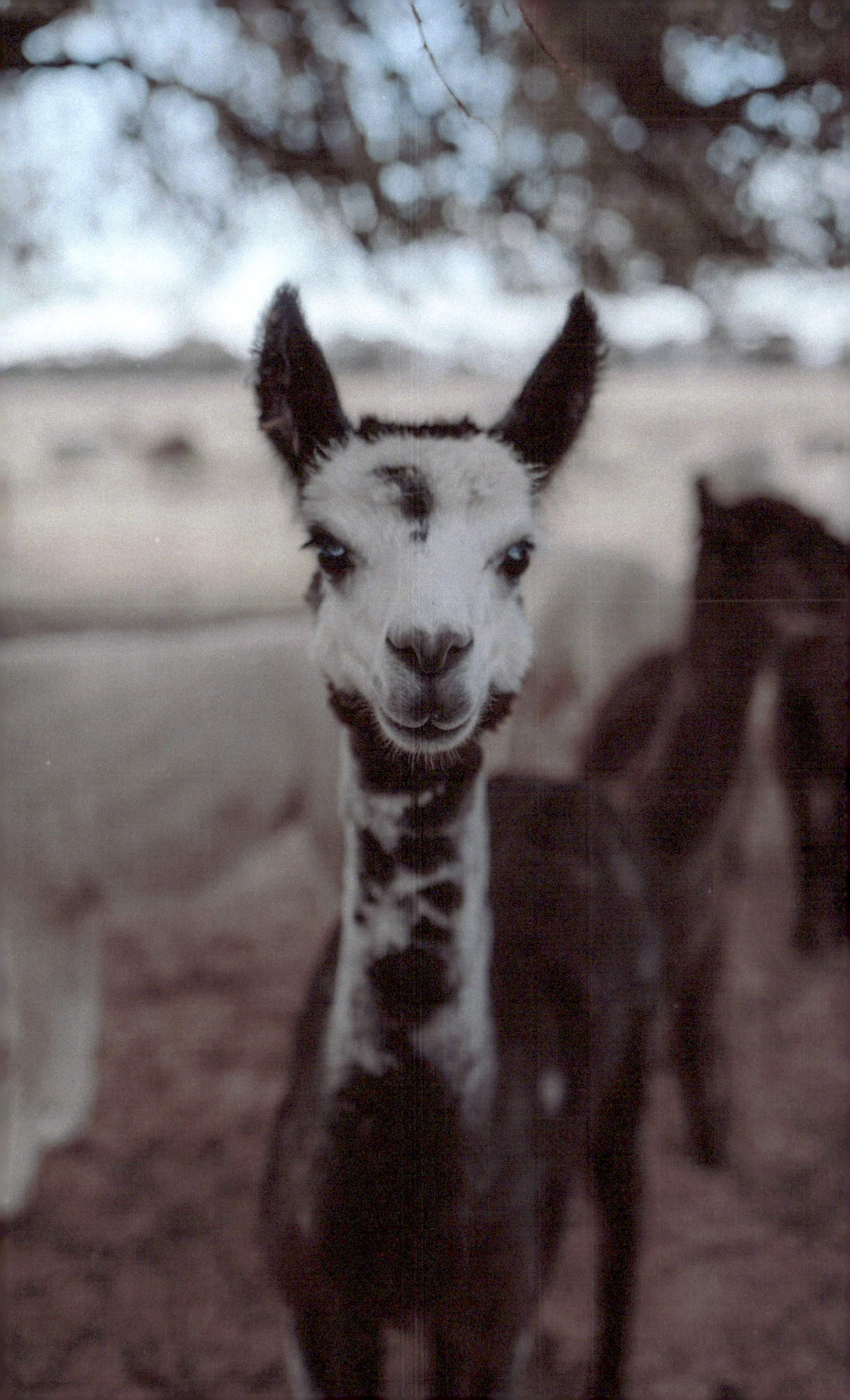

You are more than enough, exactly as you are, right now.

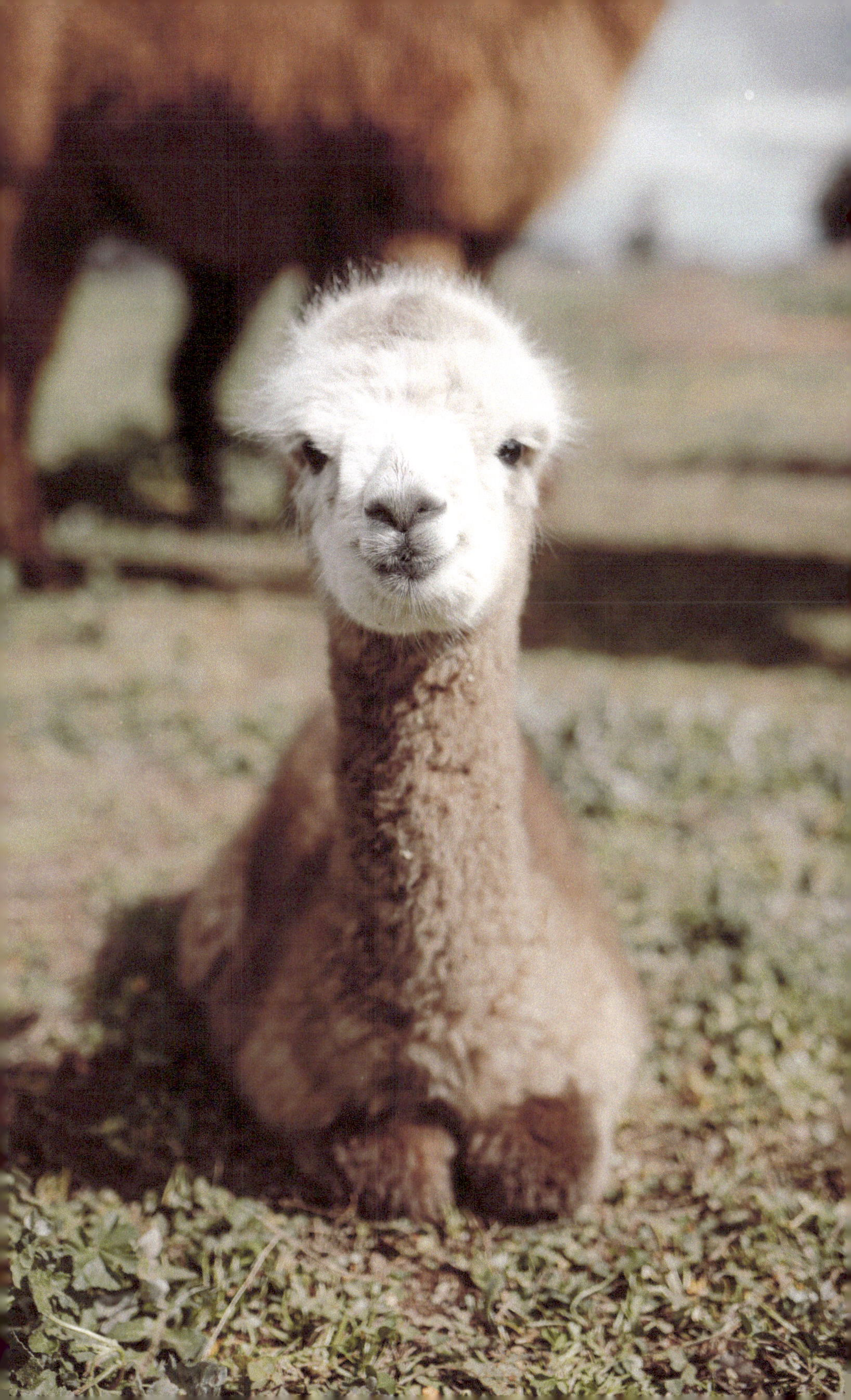

Curiosity is what turns 'what if' into something real.

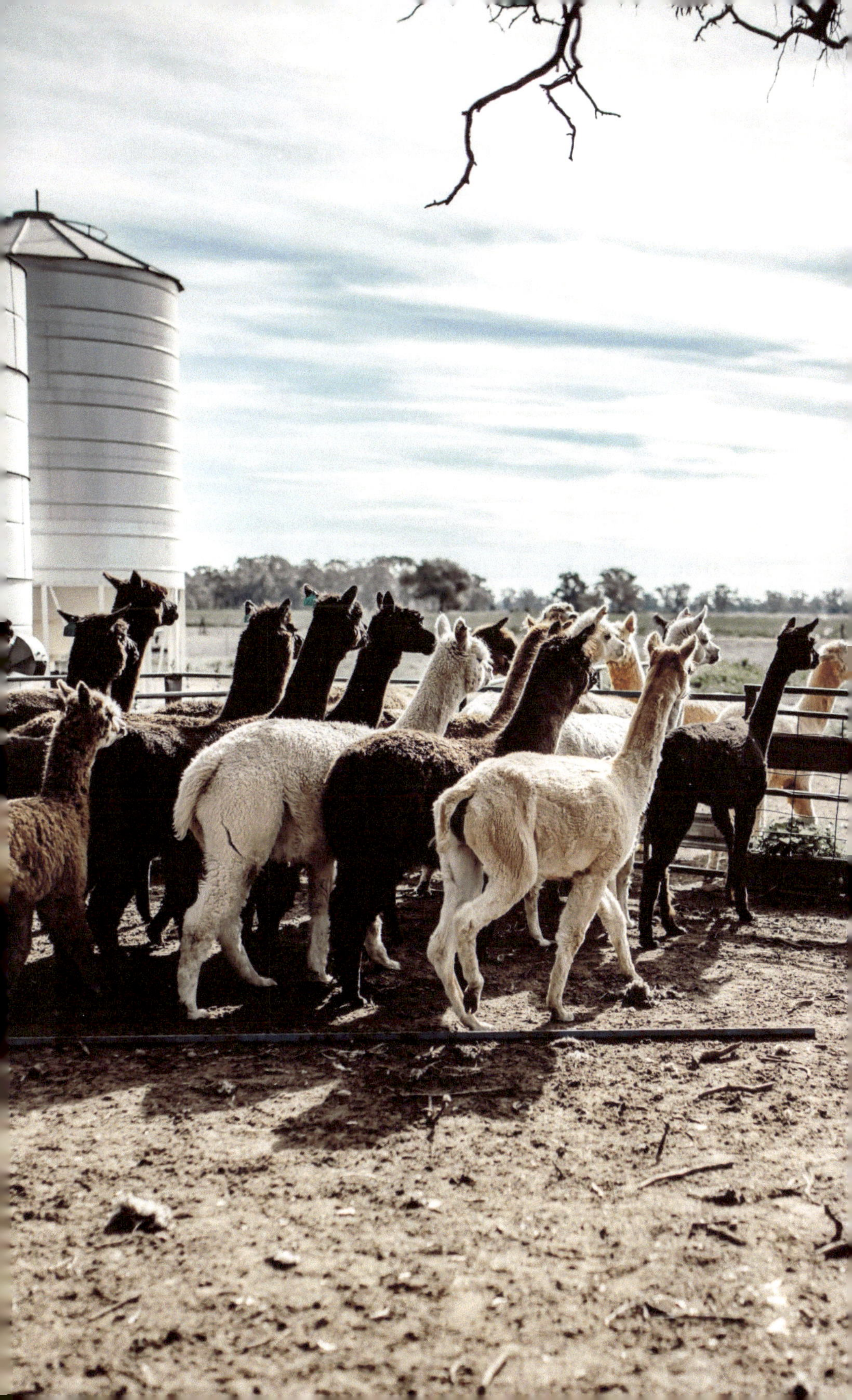

No one is too broken for a fresh start.

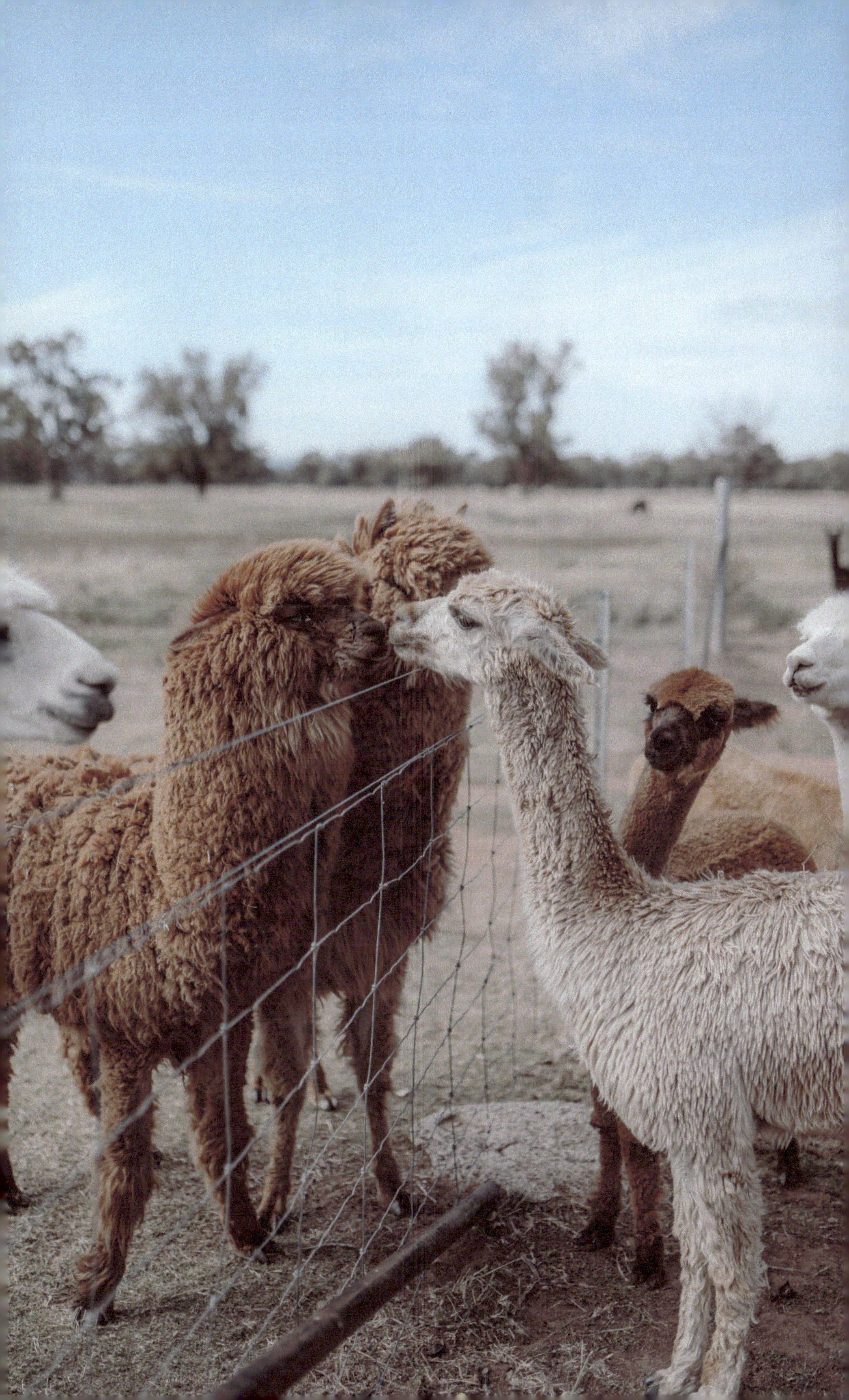

Trust that the mess is part of the making.

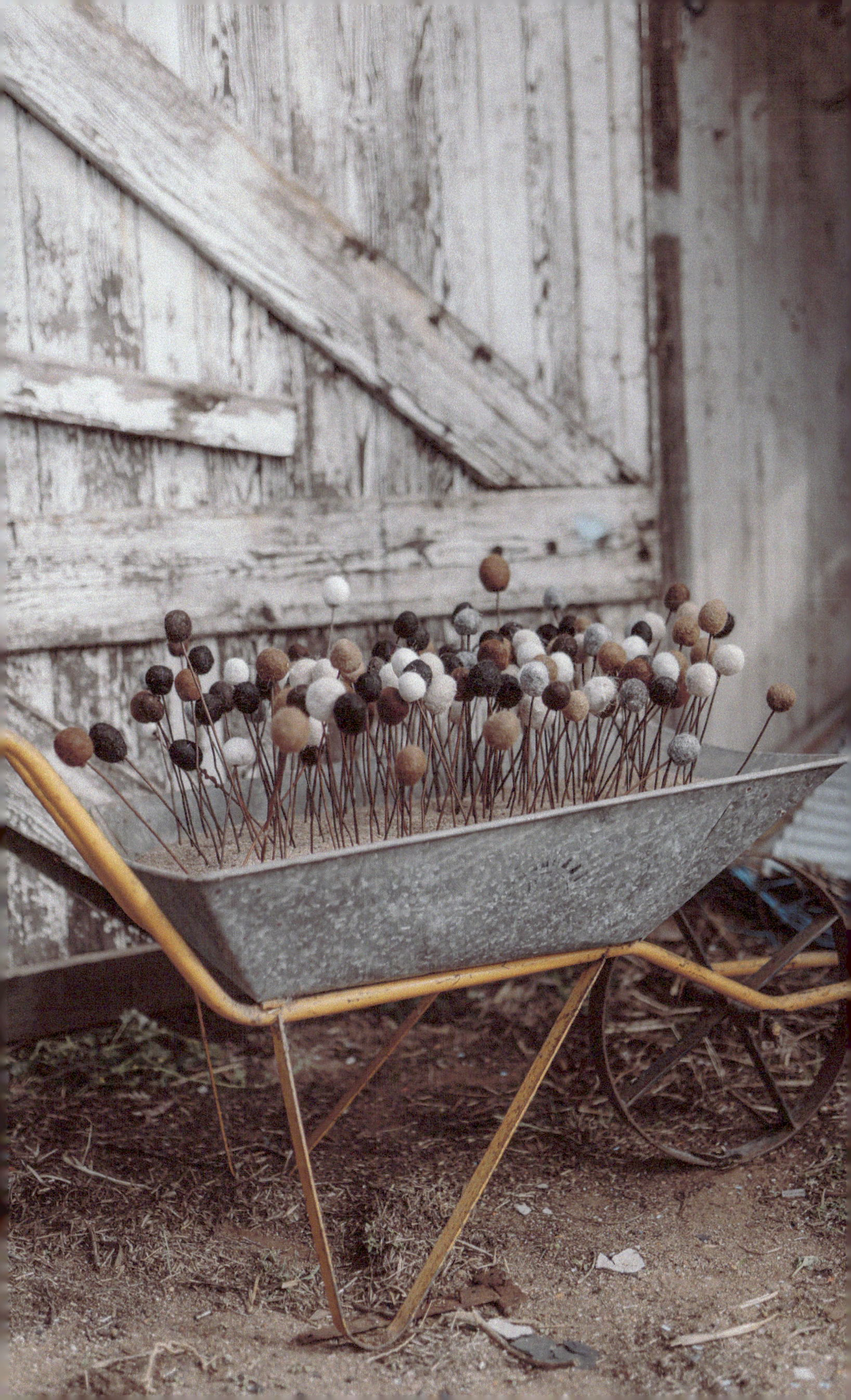

There is more than one way forward. Start where you are. The rest will meet you there.

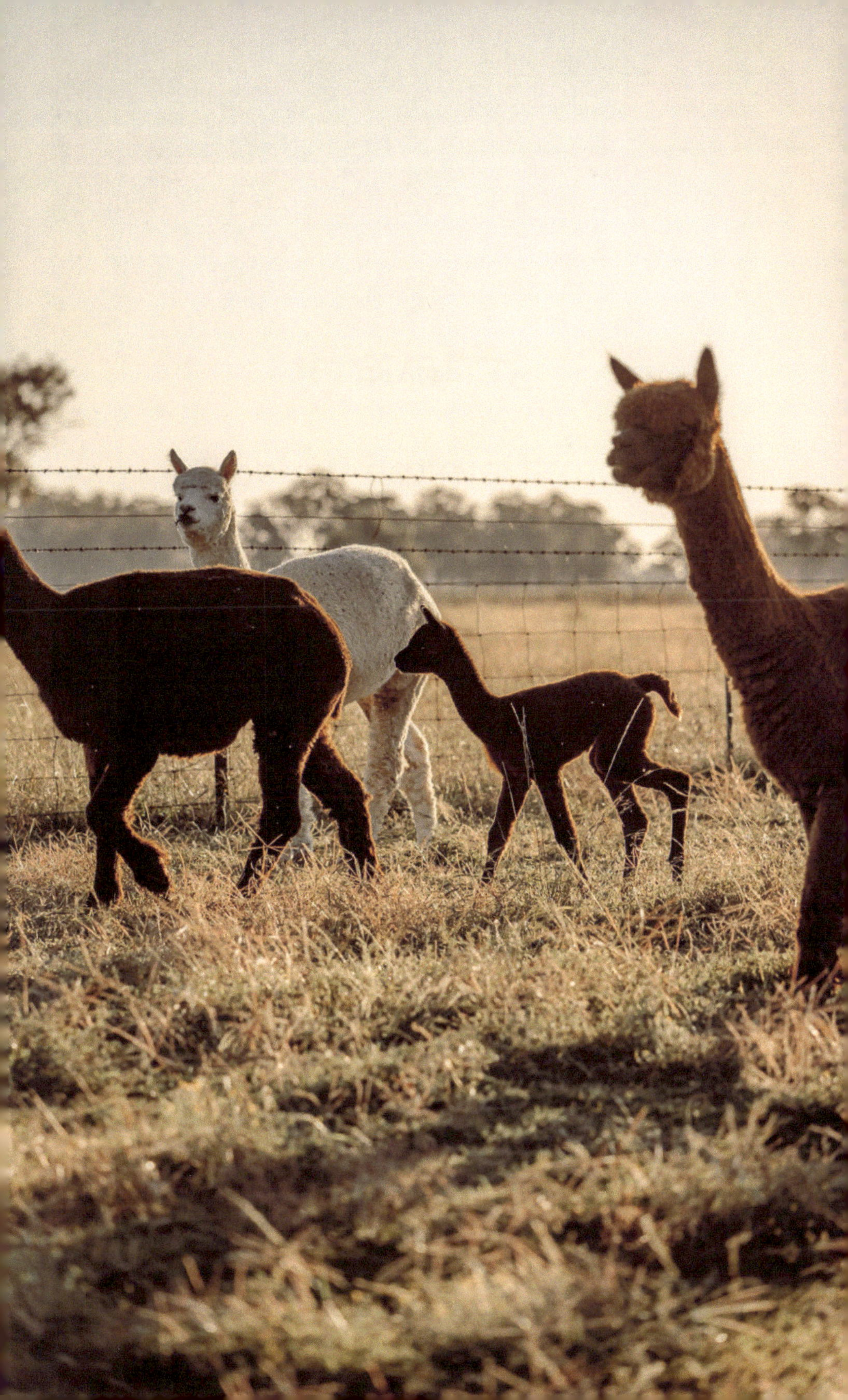

Strength

Notes from the Paddock

Strength is hard, isn't it?

There's the internal "suck-it-up, princess" kind of strength — where you take a breath and get on with it. The strength you didn't know you had until you got through a situation you'd rather not repeat. The strength that comes from knowing you've got someone in your corner.

There's strength in self-control (of which I have none), and physical strength — to fix things, build things, and get things done. Strength in numbers. And the strength to outlast a demanding toddler isn't all that different from the strength it takes to outwit a cheeky animal.

And then there's strength in conviction — the kind alpacas have in spades.

They don't second-guess which trough they're going to drink from. They drink.

They follow their instincts without hesitation and call to the herd the moment they sense a threat.

They don't change their night camp halfway through settling down — a decision is a decision.

They accept a new member into the herd — no buts, no fuss, no overthinking.

We had an alpaca called Tahlia who was the epitome of strength in conviction. An exceptional mother and steadfast herd "Aunty," Tahlia did not take kindly to anyone (especially humans) getting too close to the newest family members.

One day I was checking on one of the herds and picked up a baby to look it over. Tahlia ran from across the paddock, stopped four feet away, and coated me in spit. The serious kind. It wasn't even her baby.

I drove back to the house with all the windows down just to keep from gagging at the smell.

From that day on, she somehow confused my daughter, Emma, with me. Whenever Tahlia saw Emma simply walking between paddocks — minding her own business — she'd charge at her and spit at her. But when she saw Tyler, one of the girls who worked with us, she'd trot over for a neck rub like butter wouldn't melt.

Tahlia made no apologies for the confusion, her opinions, or her reactions.

I can assure you she didn't lose any sleep over it.

We can learn from Tahlia.

How much time do we spend — how much energy do we waste — second-guessing ourselves and replaying the "what ifs"?

How much more could we get done if we just made a choice and stuck with it?

It's not always about being sure.

It's about deciding anyway.

Make the decision and follow it through. 100%.

Life's too short for second-guessing.

It's not about never falling.

It's about getting up, dusting yourself off, and trying again.

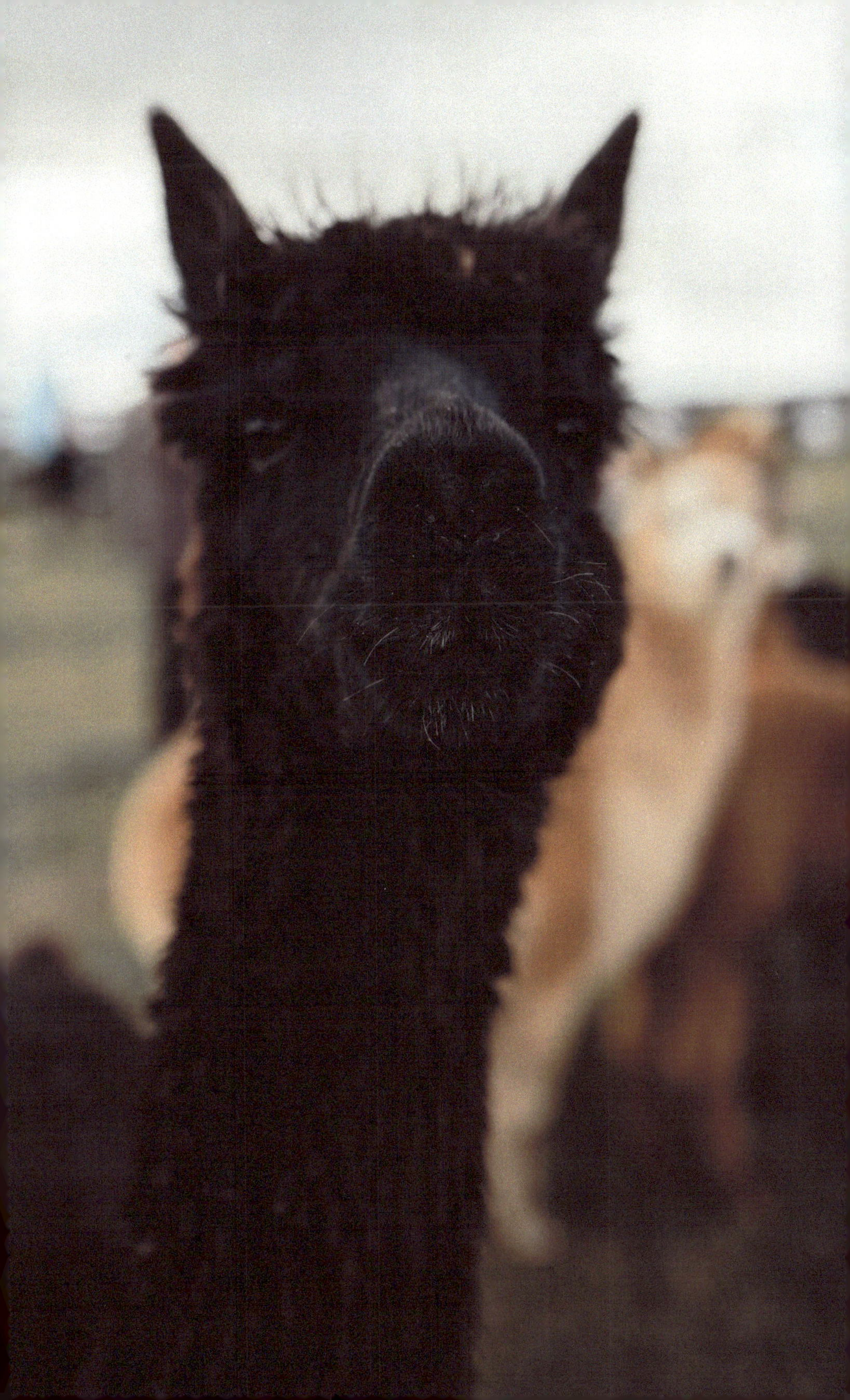

The only limits are the ones you choose to keep.

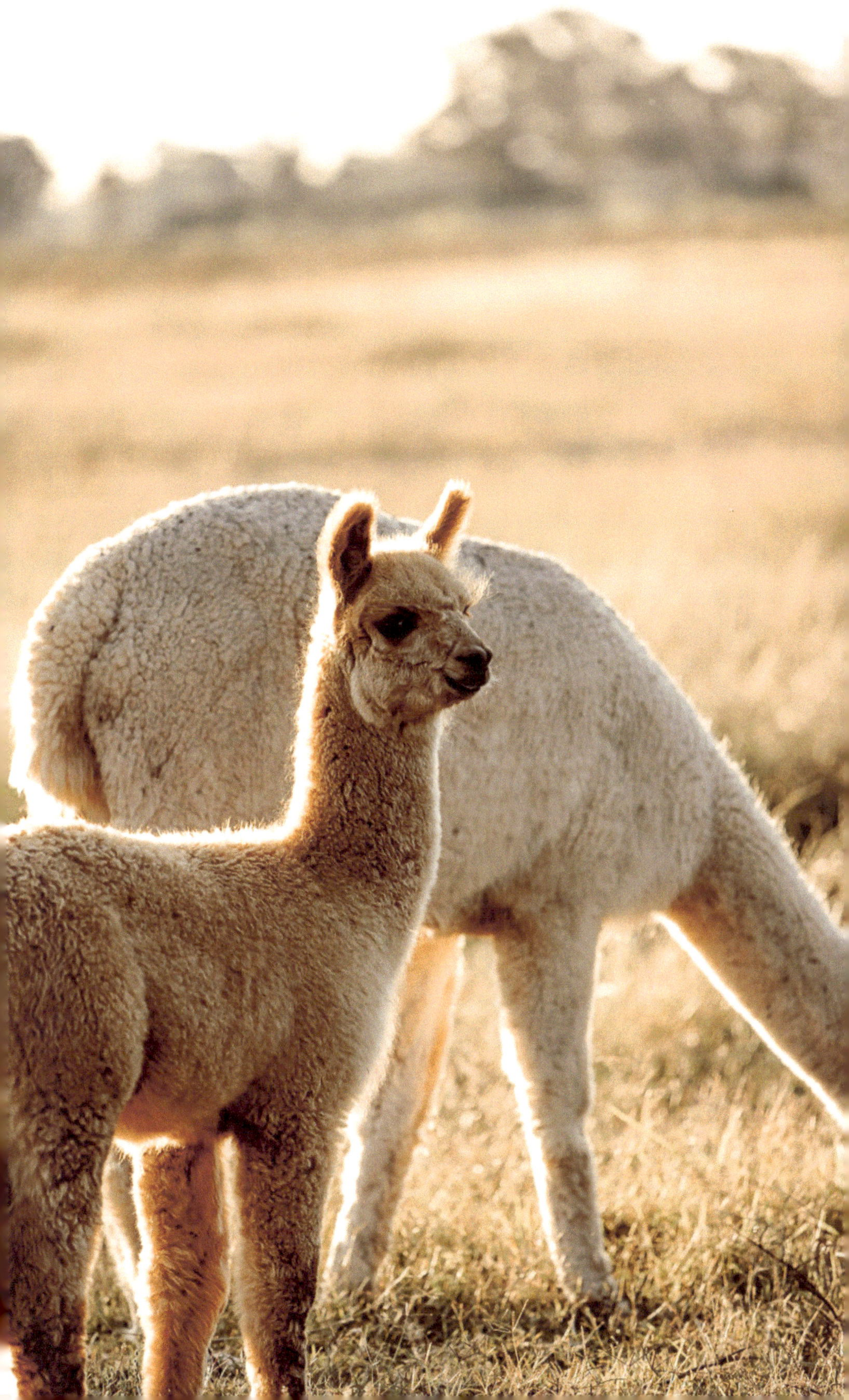

Be overwhelmed, cry, scream, stamp your feet, sulk, eat comfort food.

Now, go and do what needs to get done.

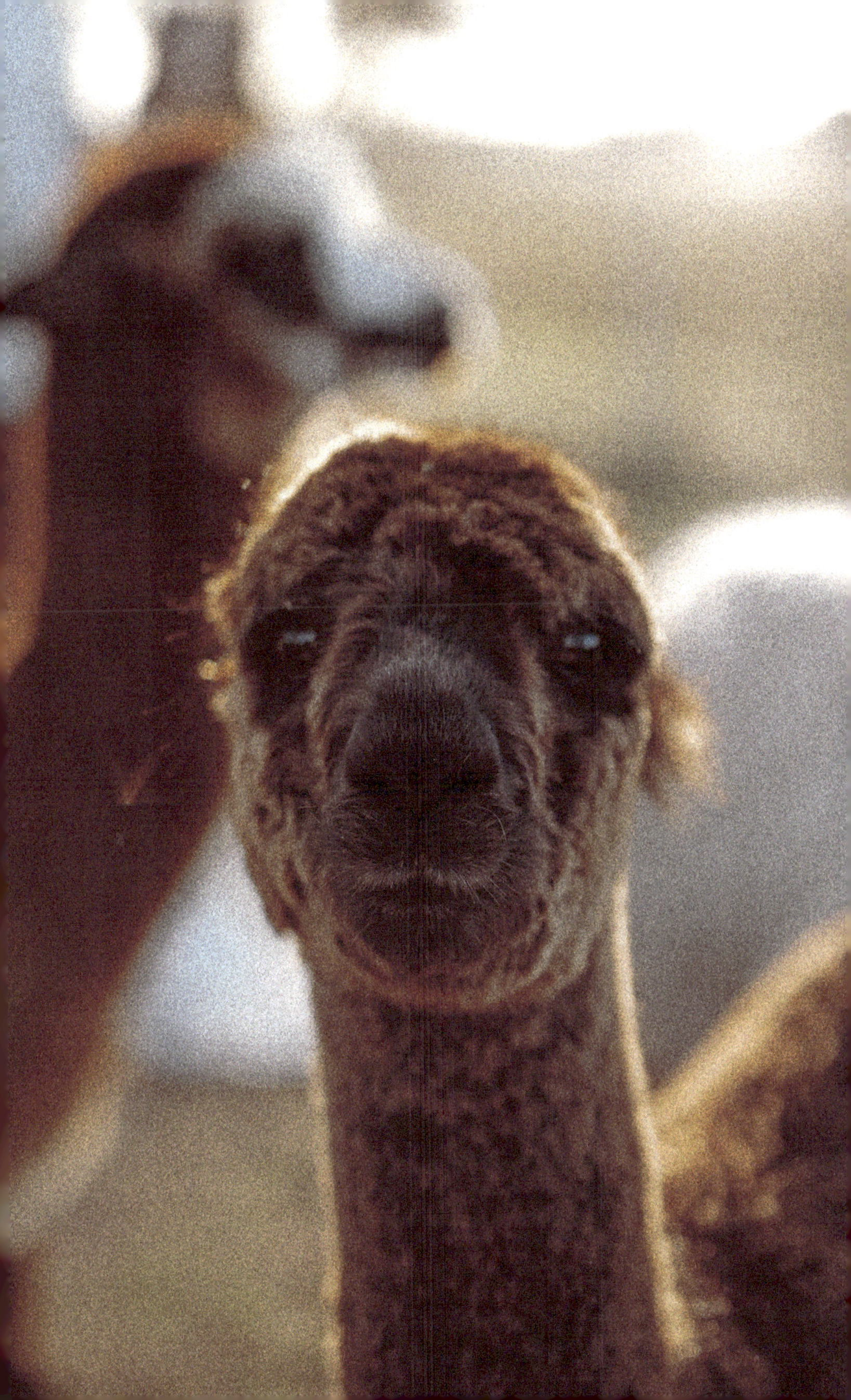

There's magic in believing you are capable of anything.

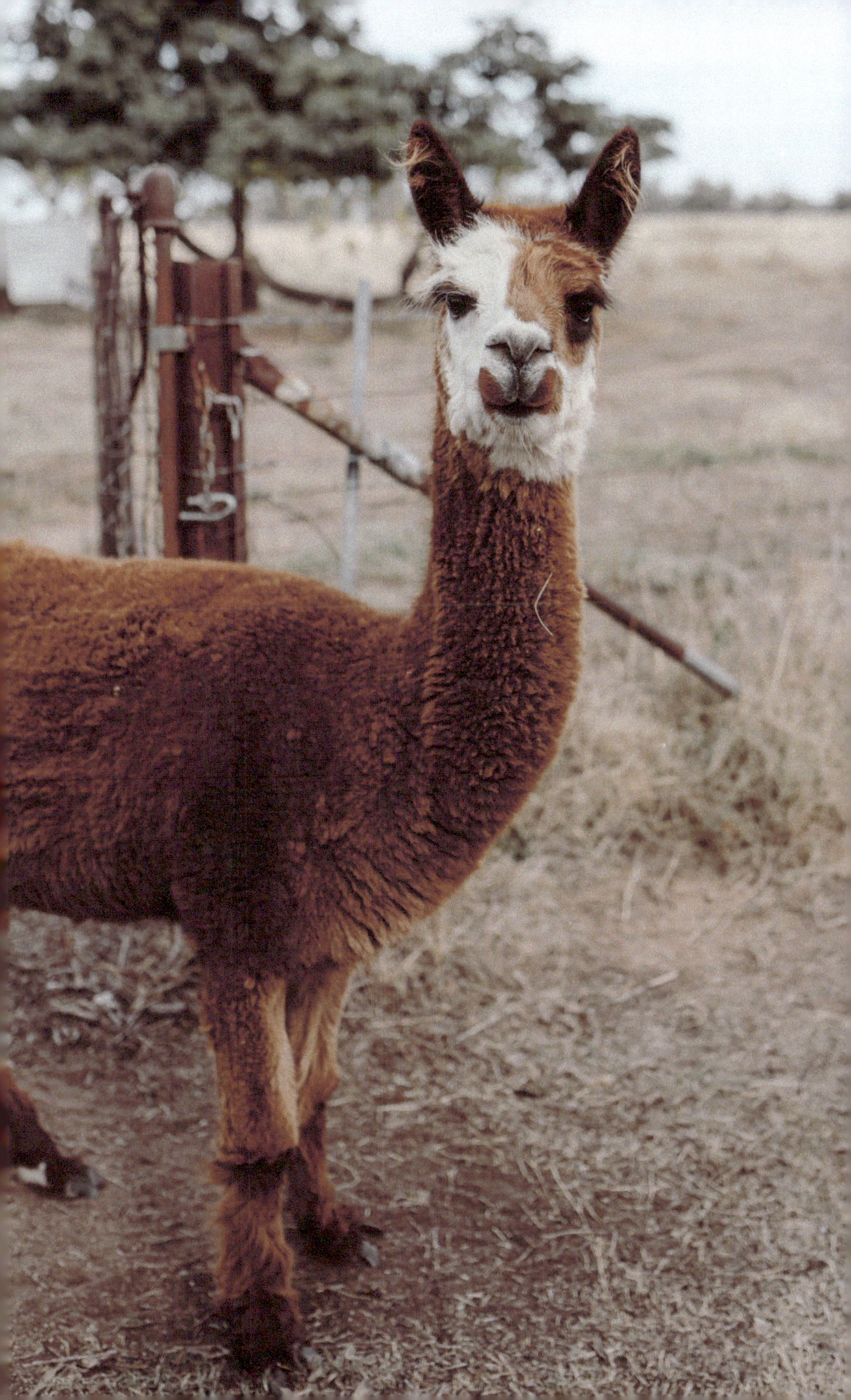

You can wait for the gate to open, or you can find your own way through.

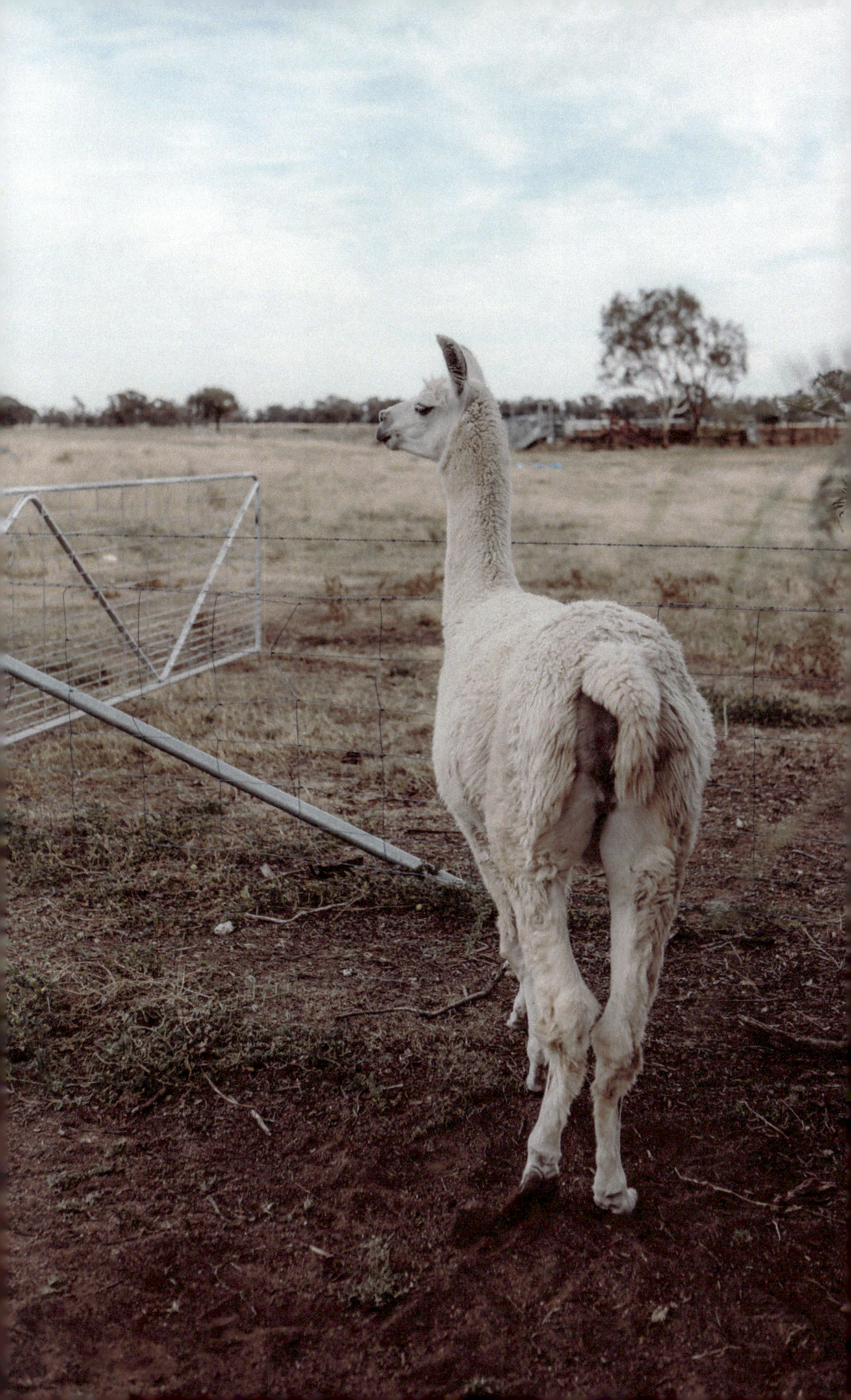

Doing what you love doesn't mean it's always easy. It means it matters.

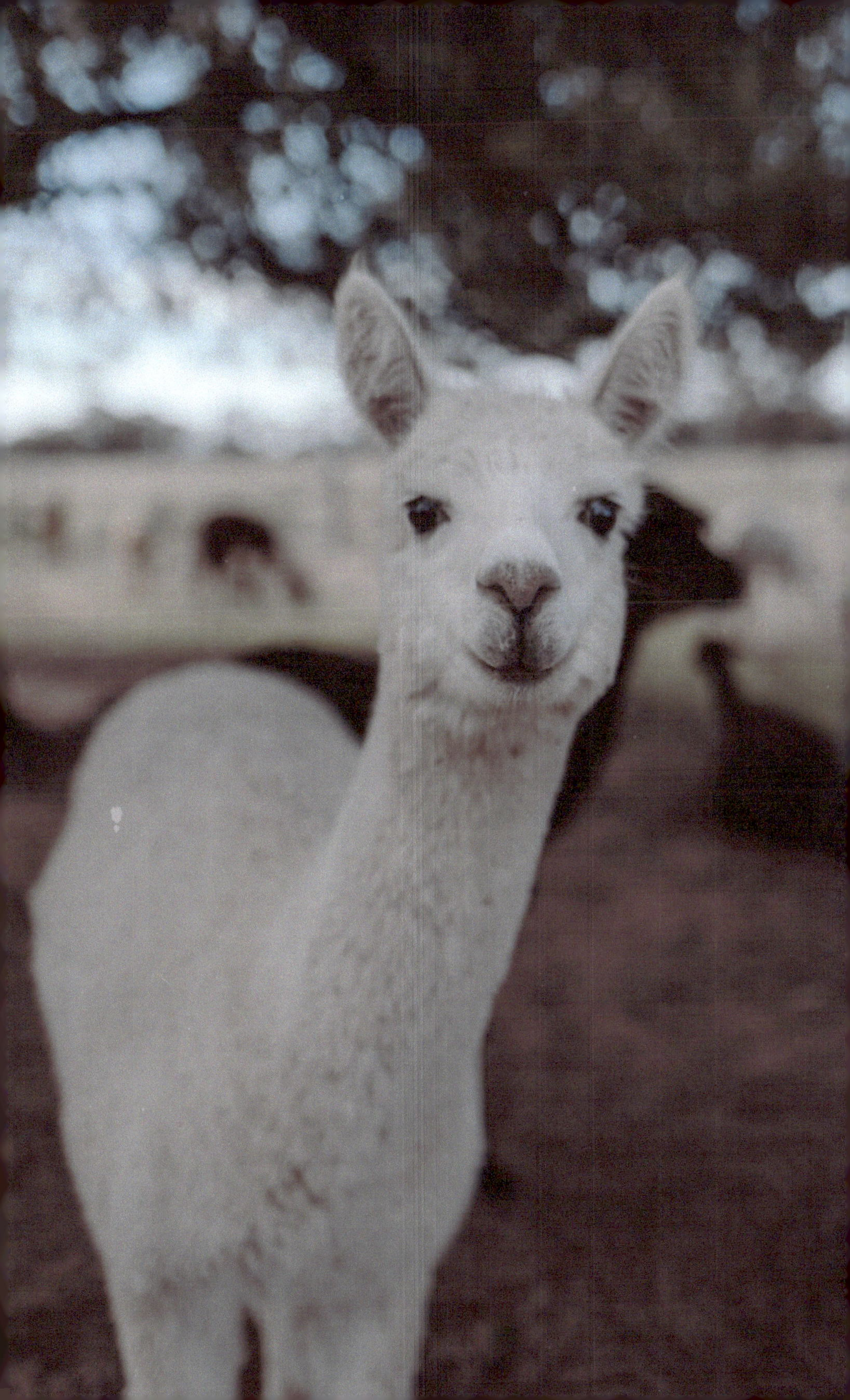

Some doors close for a reason. Some open before you think you're ready. Either way, step through.

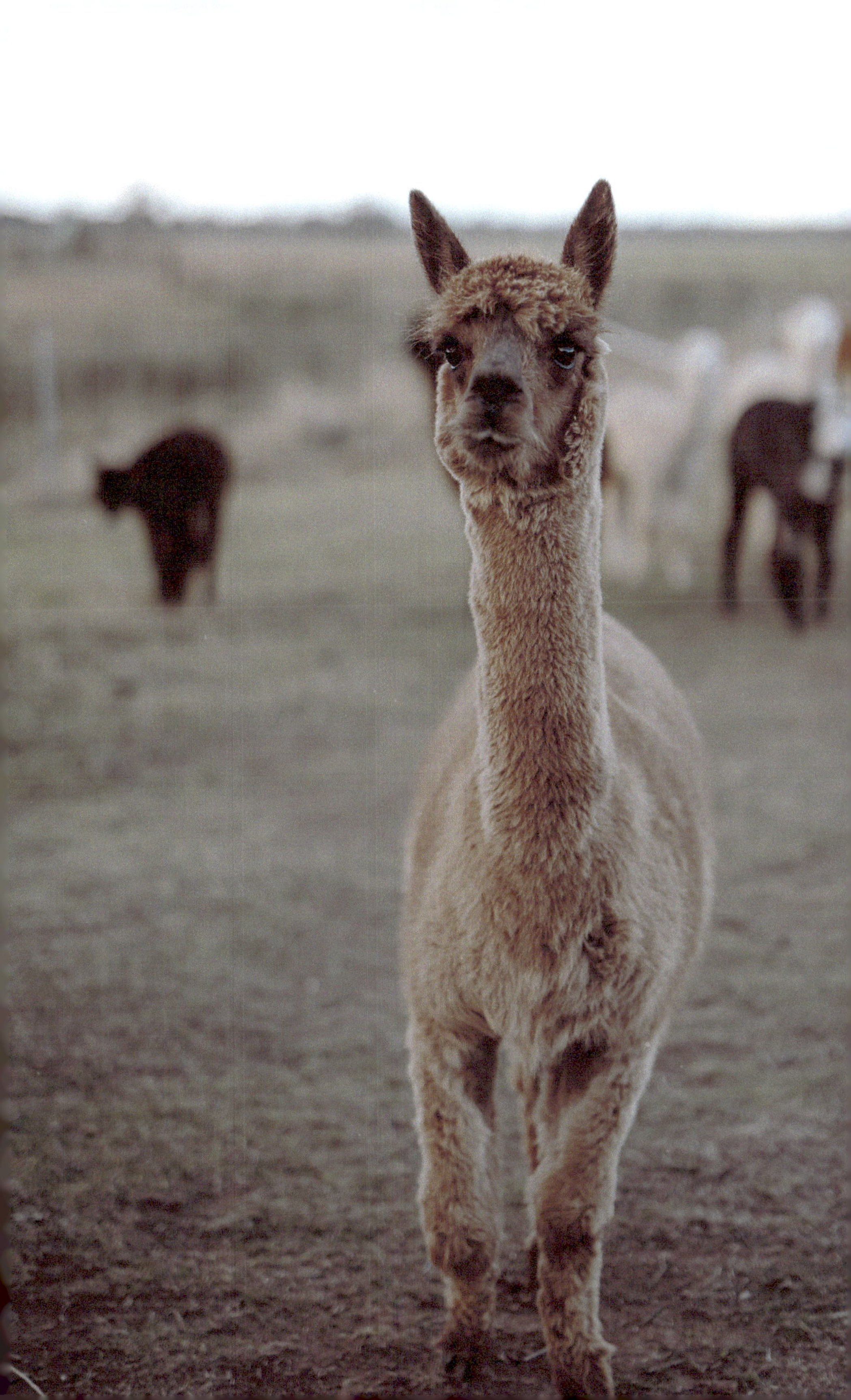

Growth comes not from what you've lost, but what you choose to rebuild.

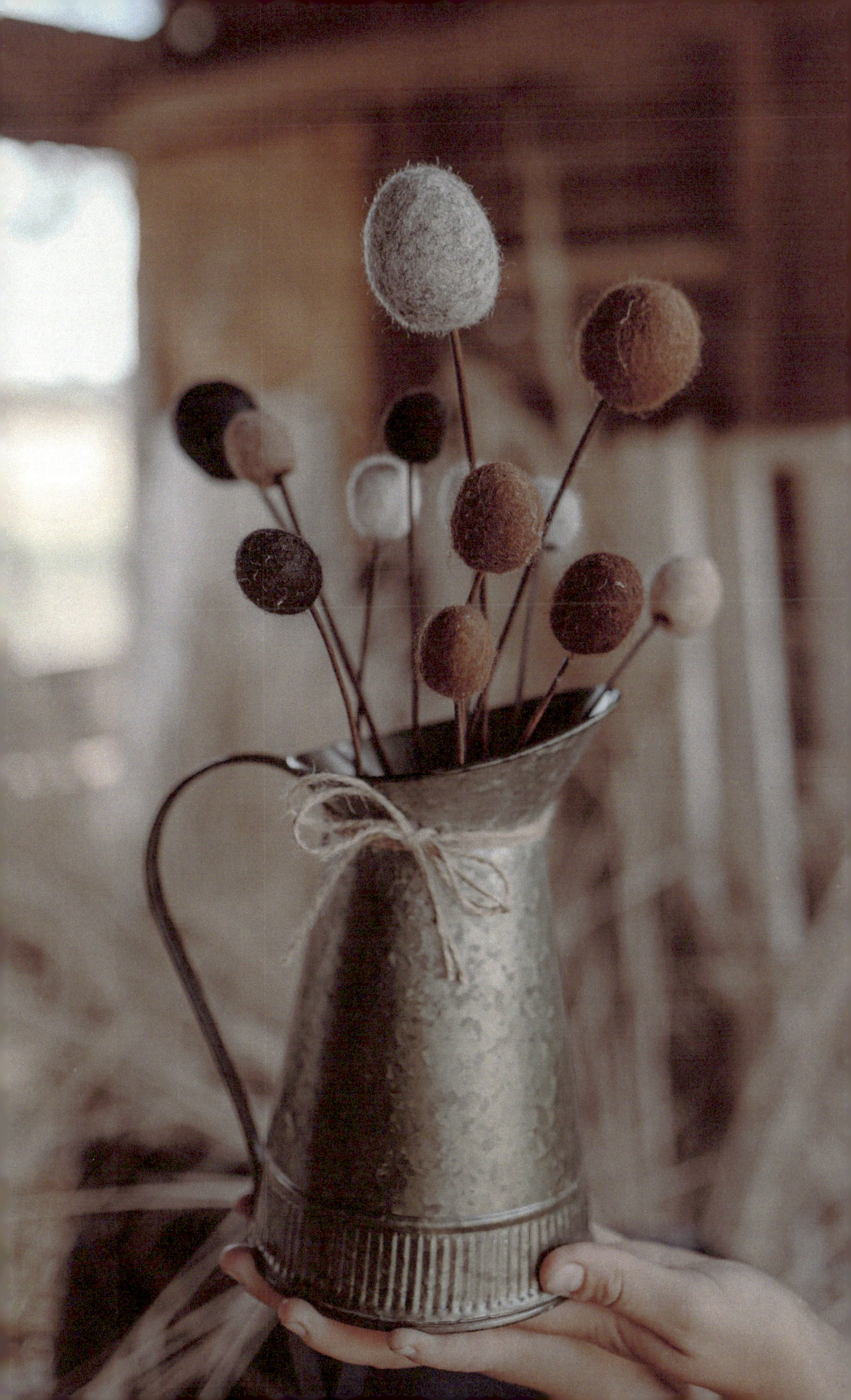

Some days, survival is success.

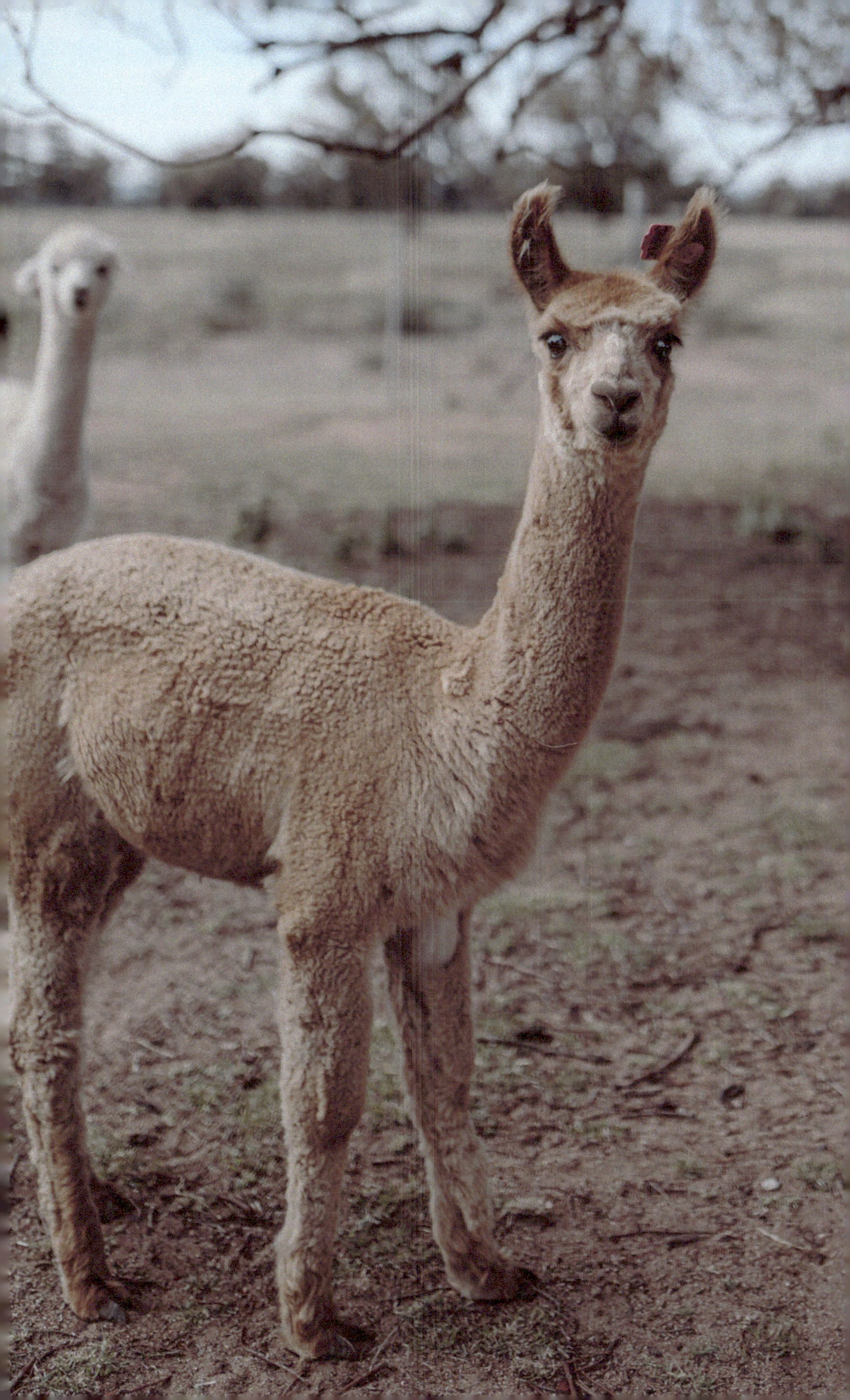

The best story is written when you are brave enough to live it.

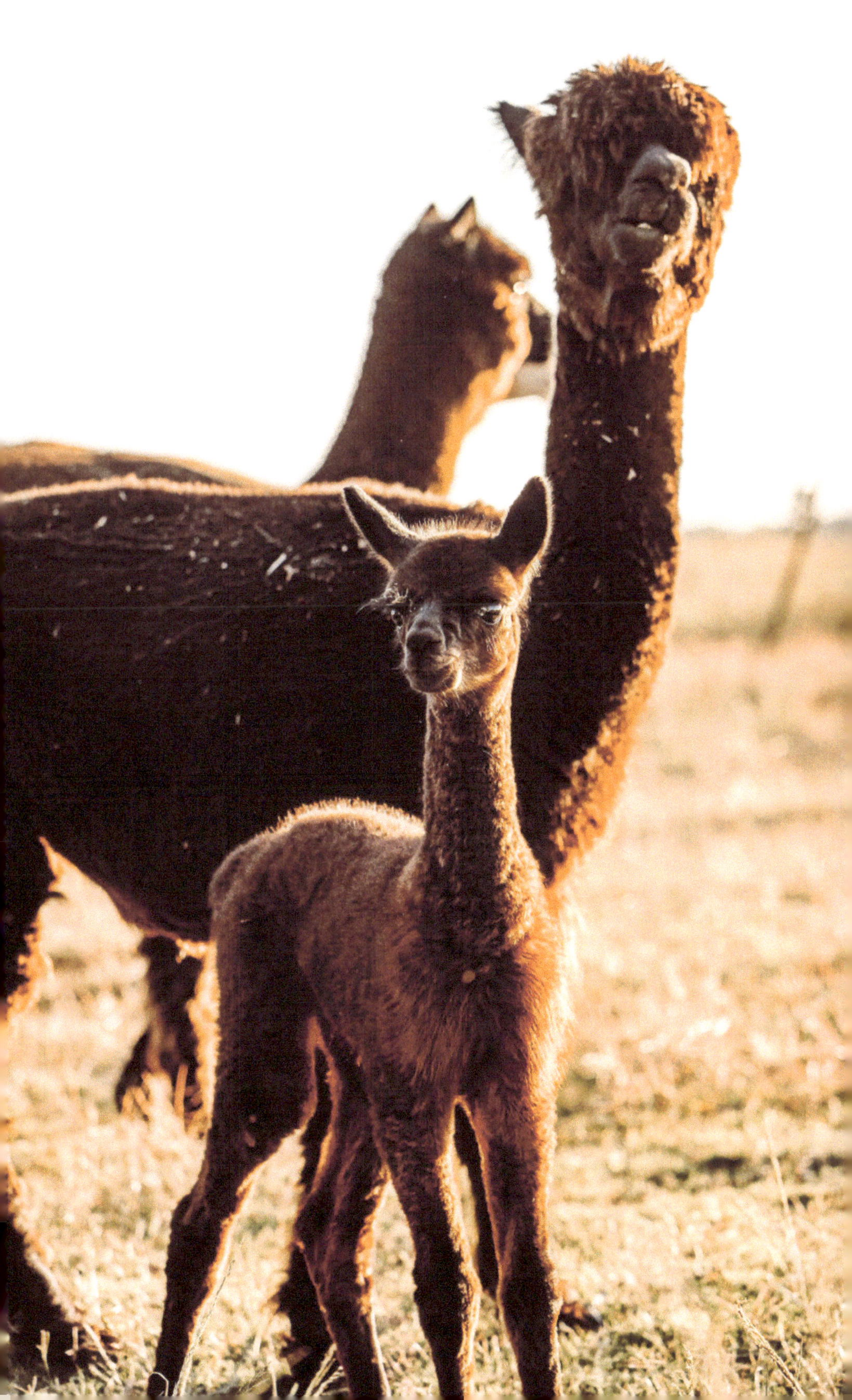

Authenticity

Notes from the Paddock

Alpacas are unequivocally, unapologetically themselves.

They don't perform, pretend, add a filter or polish the edges — they just are.

Robyn and her young daughter Bobby came to live with us early on.

Robyn took to farm life like a duck to water — quickly learning that human visitors meant food. She was always the first to the fence, cheeks full of pellets, happily invading anyone's personal space for a neck rub.

We were a little puzzled that Robyn never brought Bobby with her.

She always hovered in the background, and we assumed she was just naturally shy — the quiet one, the opposite of her mother.

Then came weaning day — the time when teenage alpacas move out of home, away from their mothers, and learn to be big kids. It's always a tough week. Someone's always crying at the gate — sometimes the mother, sometimes the baby. We usually take bets on who'll sulk the longest.

The mothers were moved into one paddock, the weanlings into another.

Bobby raced full-speed along the fence line before leading the others in a wild pronking session — springing and leaping around the paddock like she'd been waiting her whole life for that moment.

We stood watching, laughing until our sides hurt.

Turns out it wasn't Bobby who was shy at all. For reasons known only to her, Robyn had been holding her back. Once Bobby finally had the space to be herself, she took it — and took us along for the ride. It was beautiful.

We'd misunderstood her completely — not because she was hiding, but because we weren't looking closely enough.

Turns out, being yourself isn't something you learn.

You just need the room to do it.

Beauty isn't in perfection. It's in authenticity.

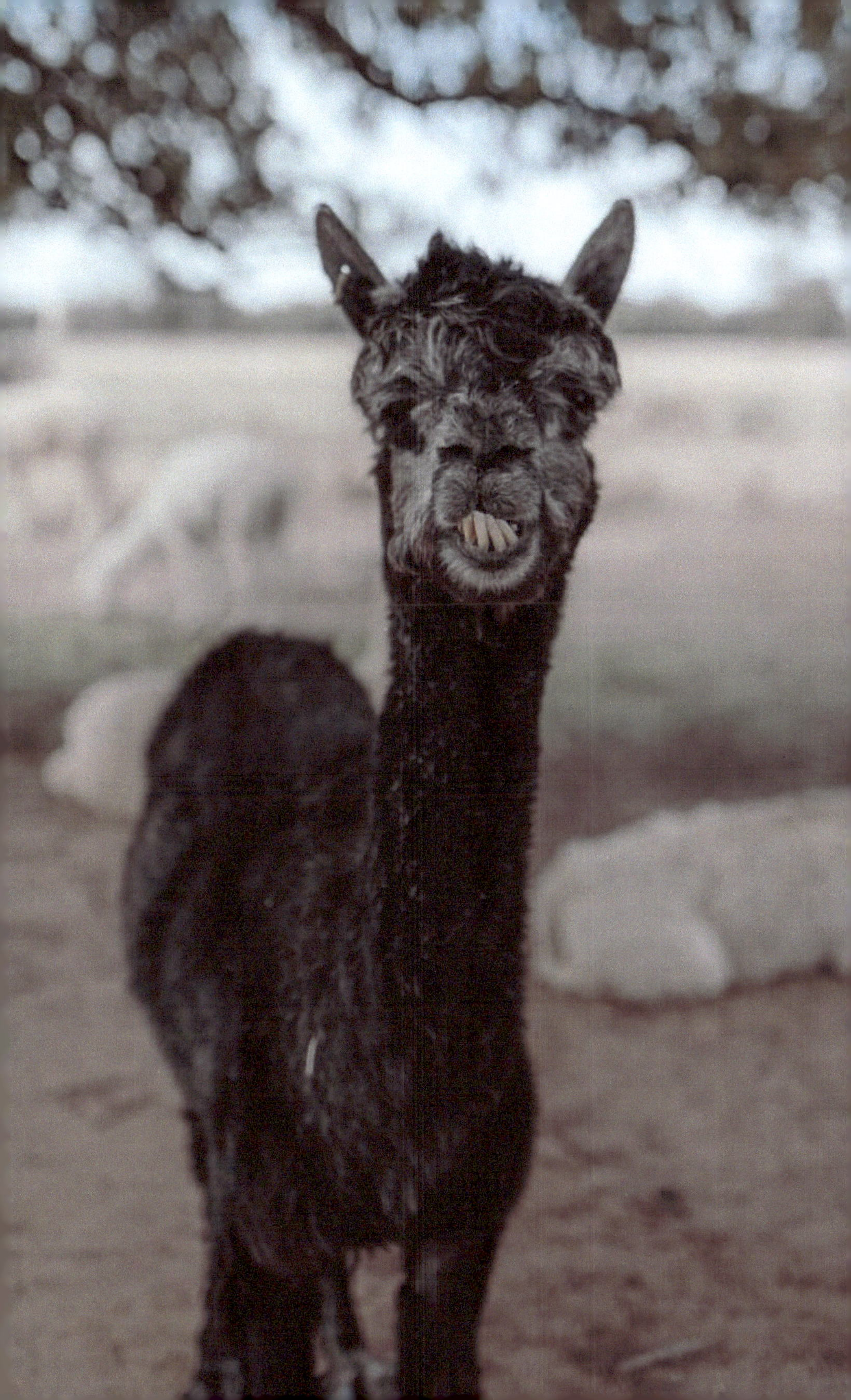

You don't have to stand in the spotlight to shine.

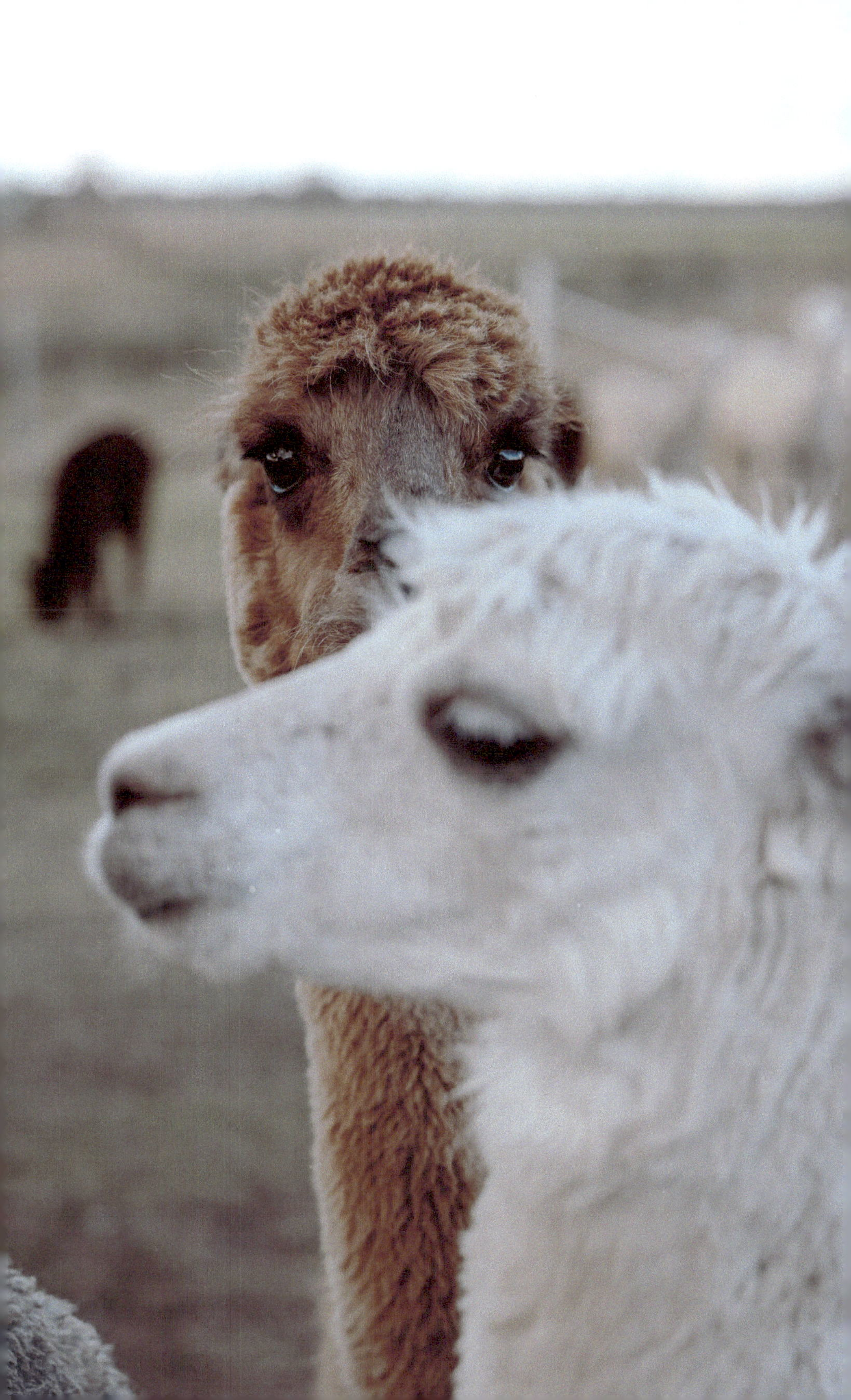

Find beauty in the imperfect, strength in the unfinished, and thrill in the unknown.

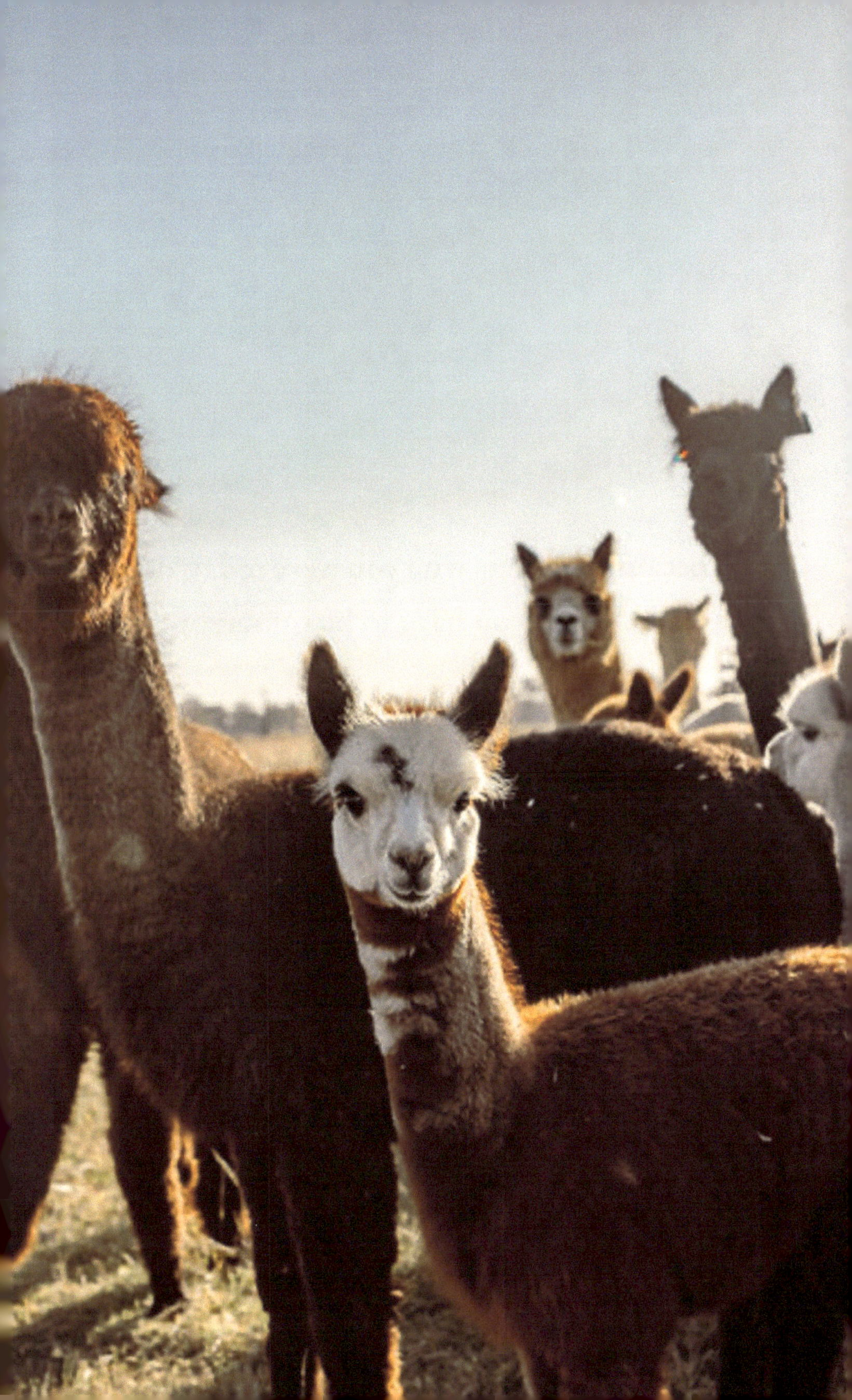

Just because that was who you were today, doesn't mean that's who you have to be tomorrow.

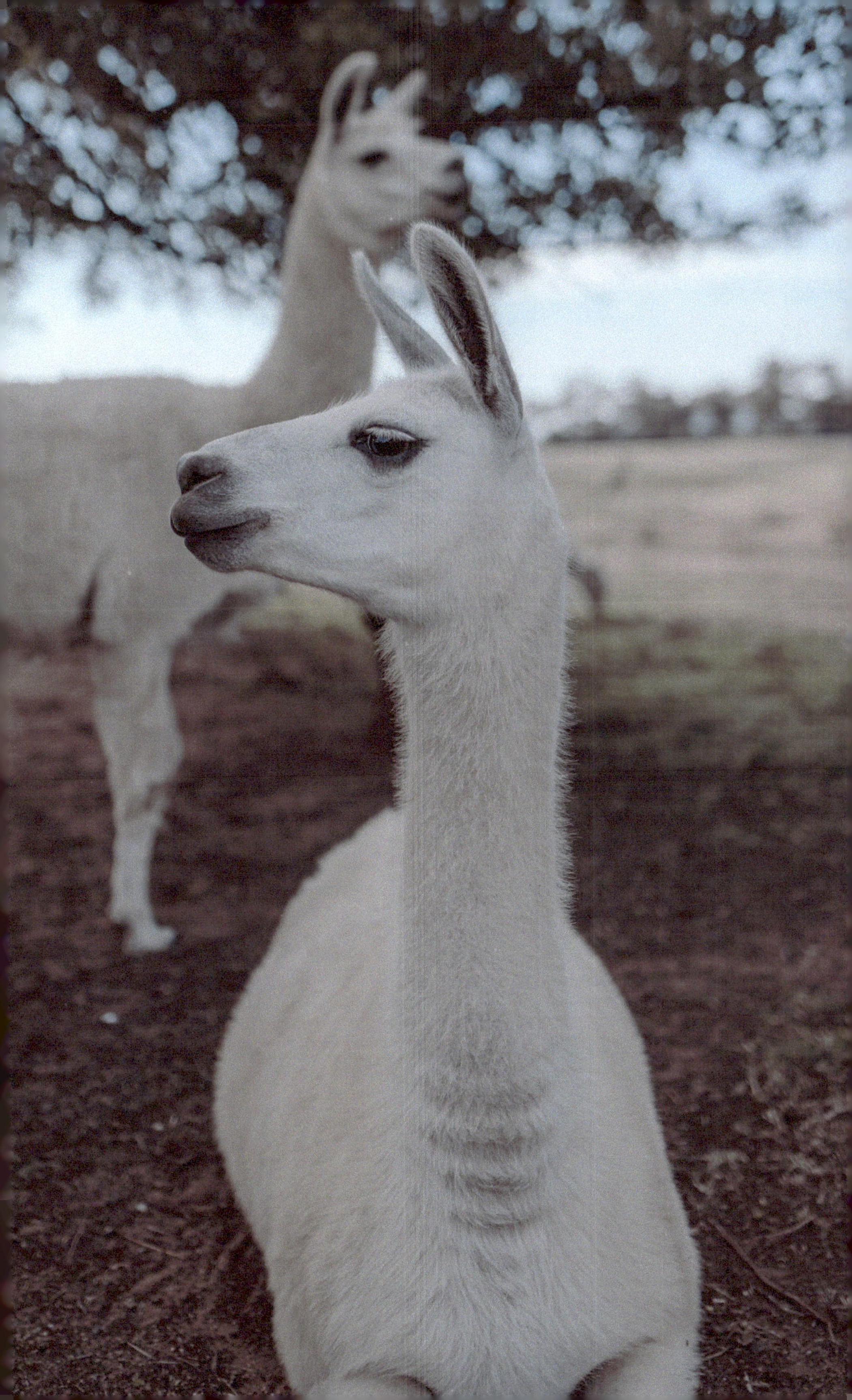

The world needs more of what only you can give.

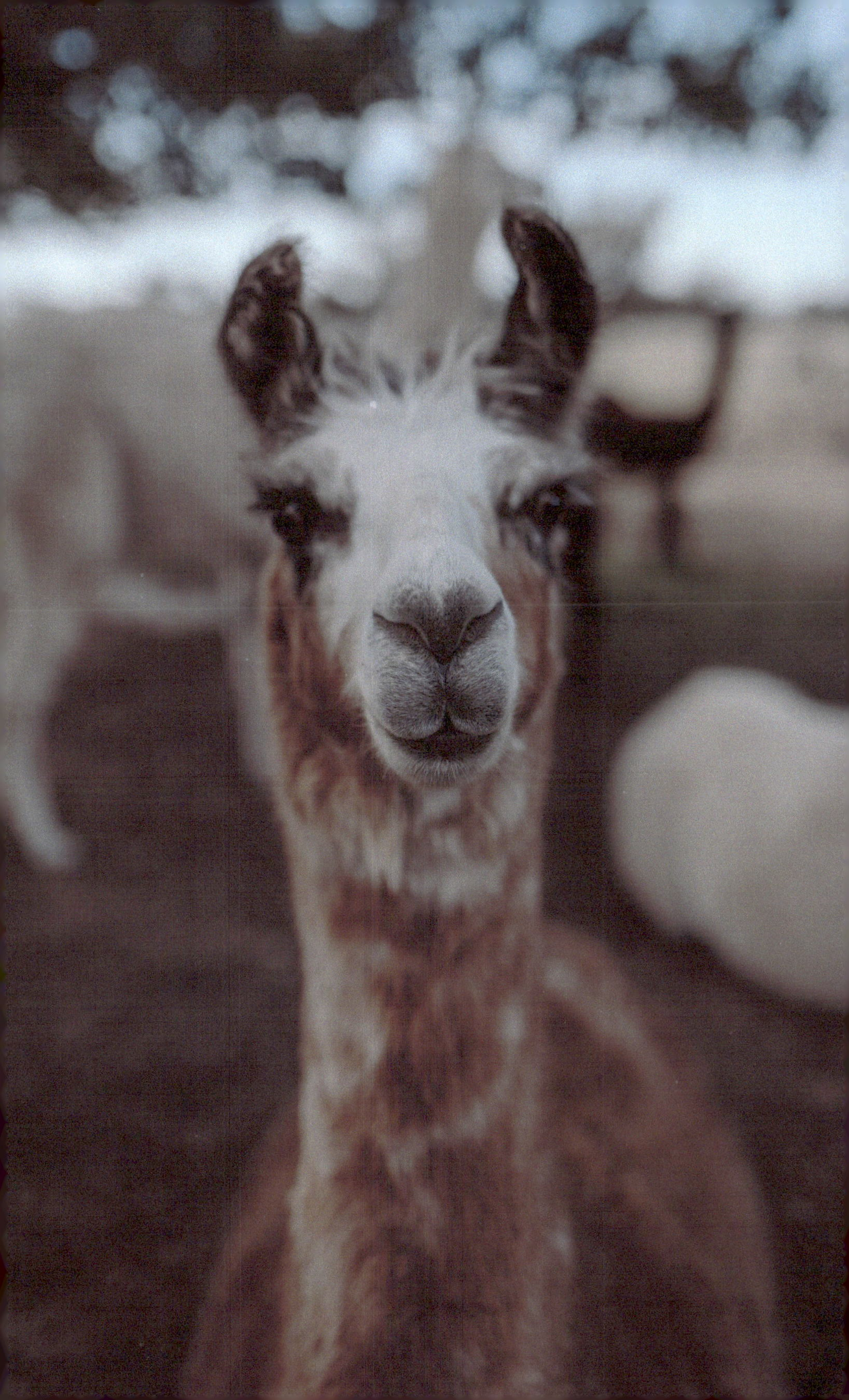

Loving what you do doesn't mean loving every moment of it. It means knowing, deep down, that you wouldn't trade it for anything else.

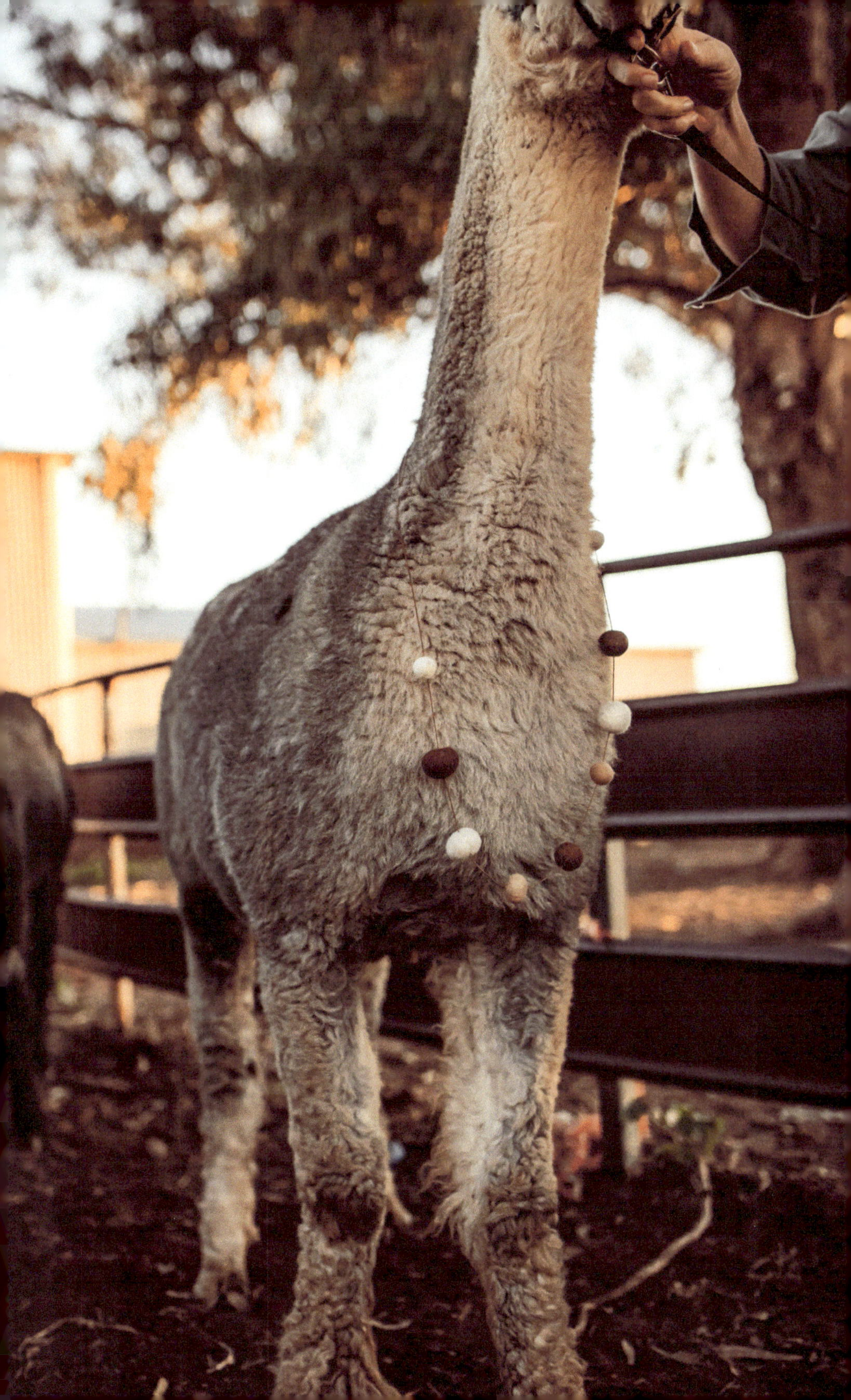

Some days, you're juggling. Other days, you're balancing. Both are okay.

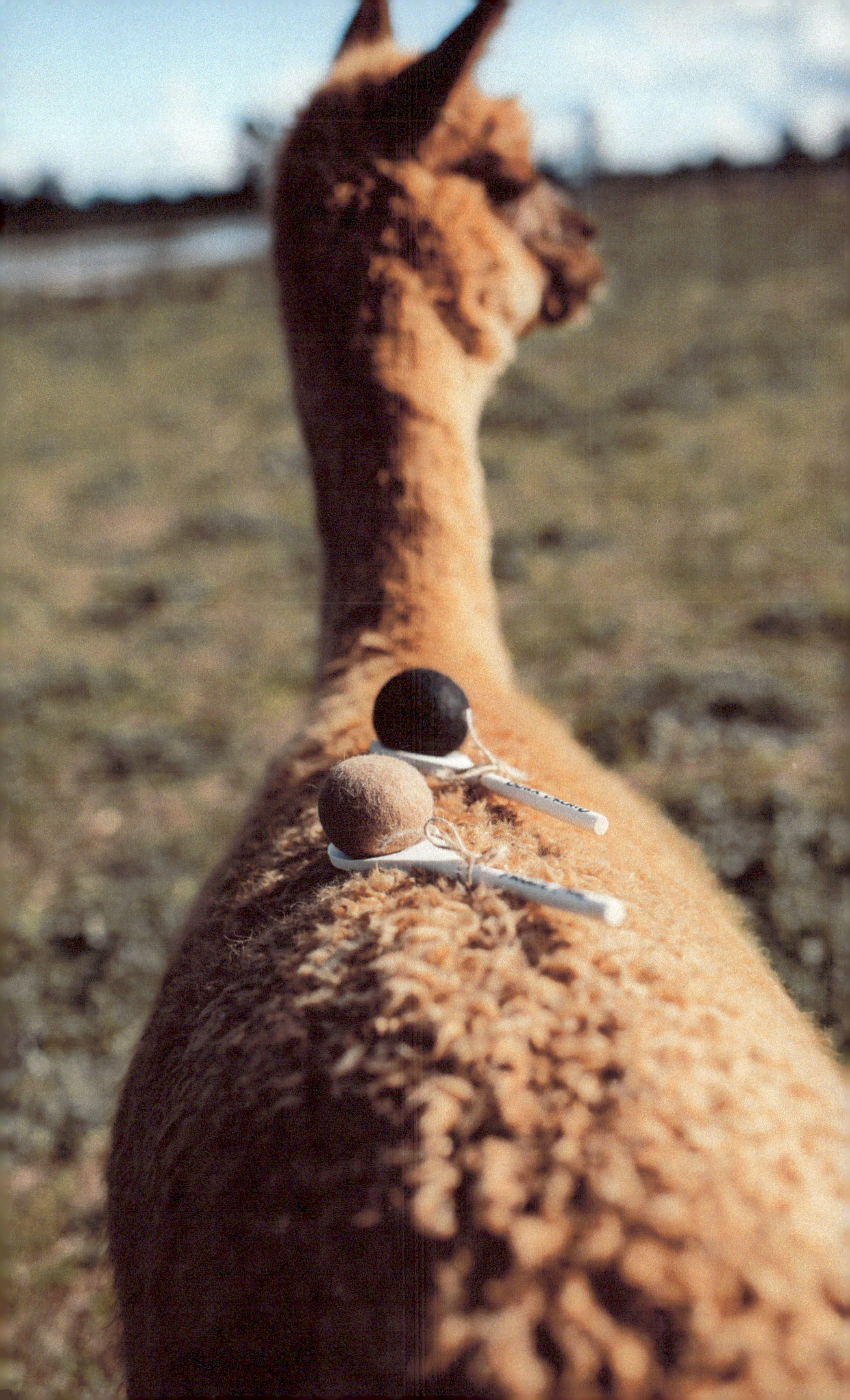

Celebrate the little things. They're the big things in disguise.

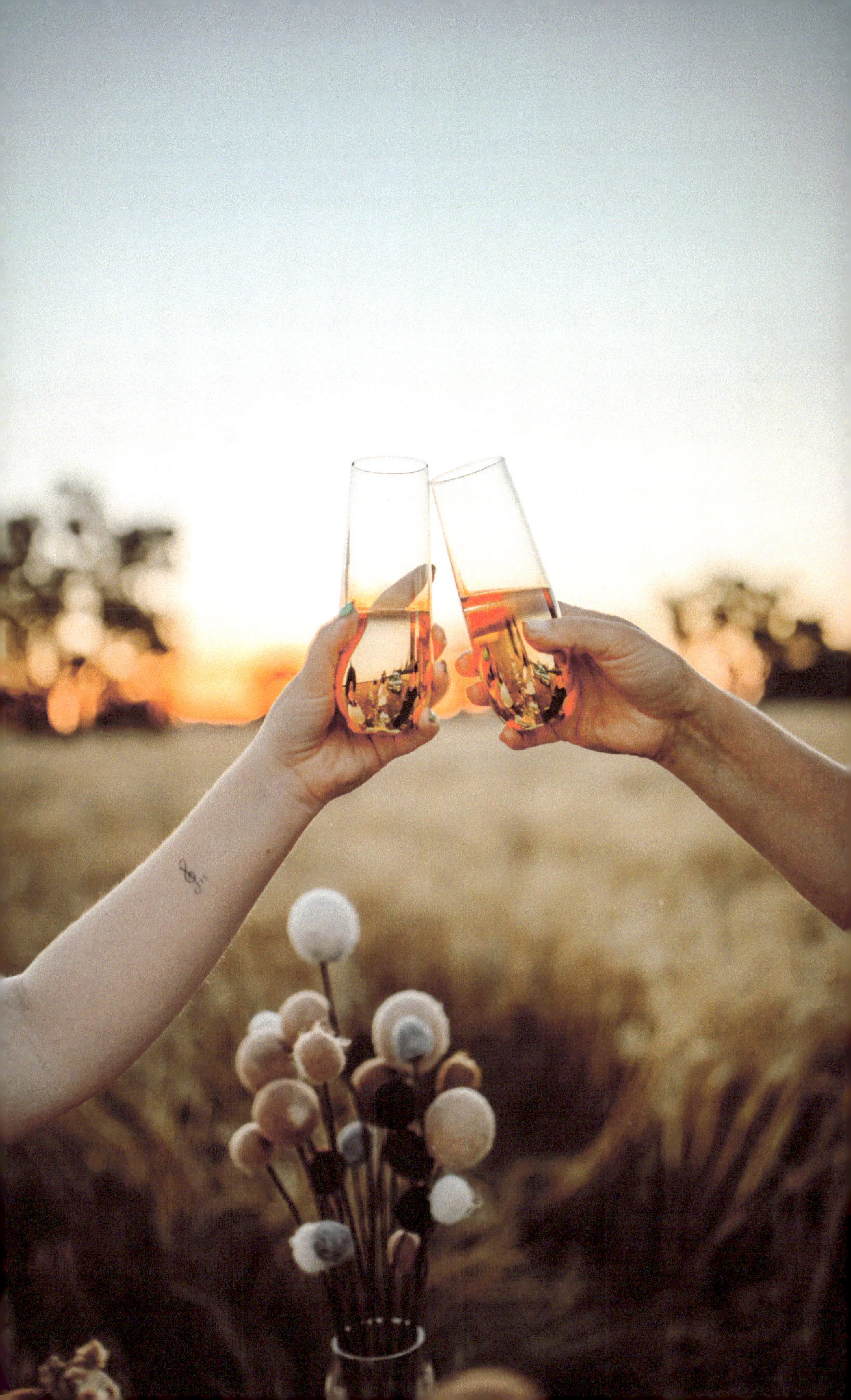

Not everything needs fixing. Some things just need noticing.

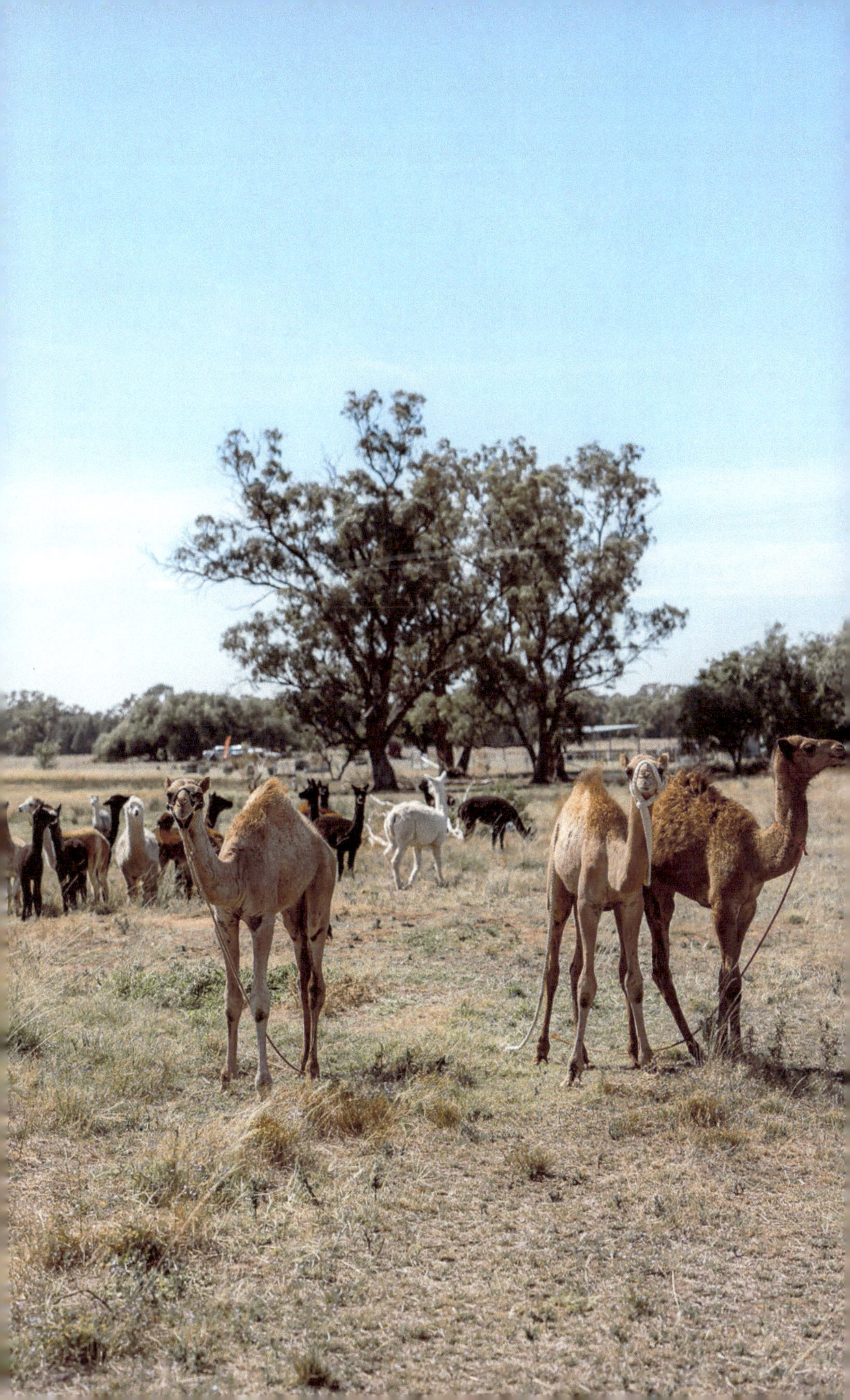

Sometimes, following your heart doesn't make sense. Until one day, it does.

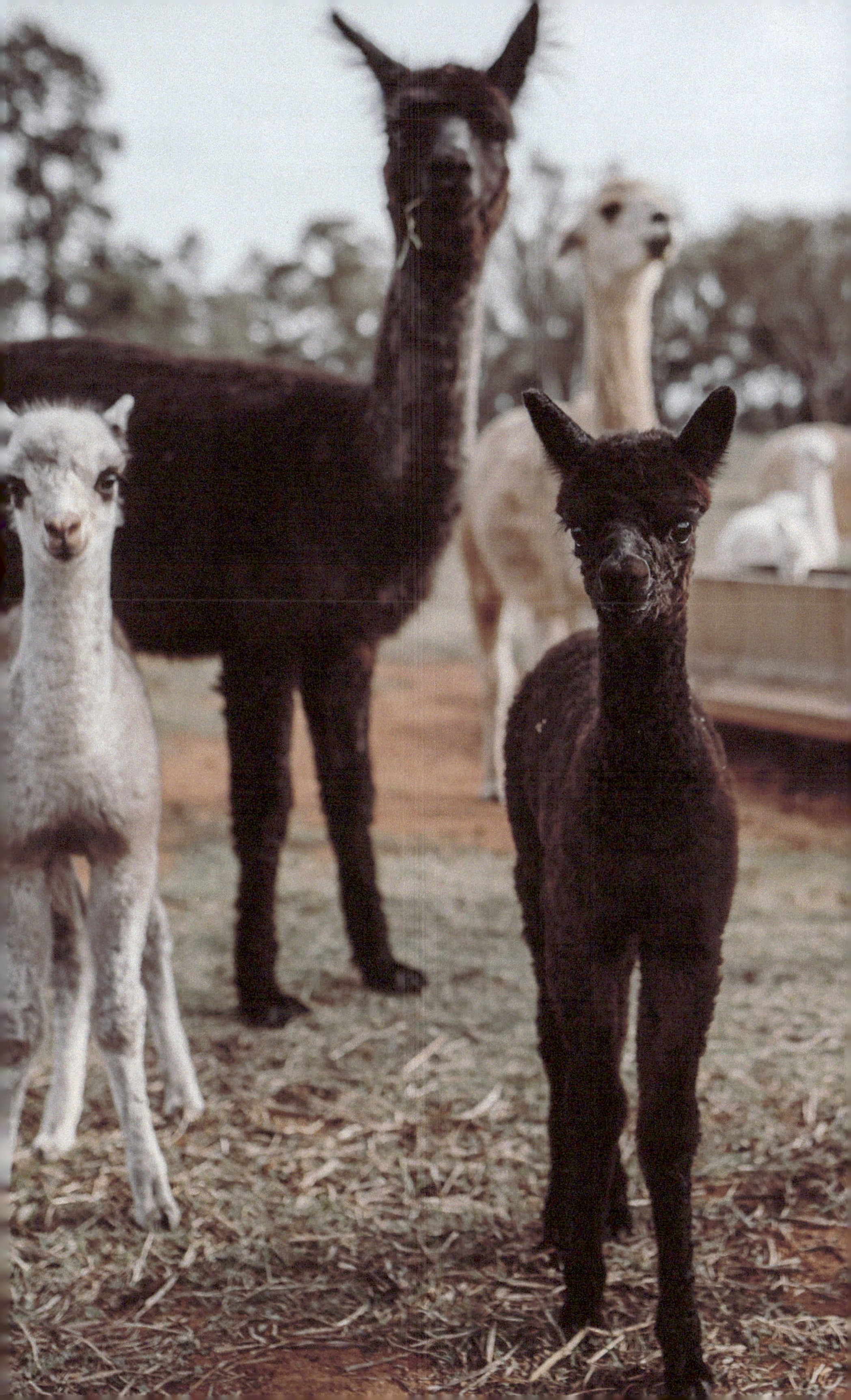

CONNECTION

Notes from the Paddock

We can't do it alone.

Over the years, we've taken in more than sixty rescue alpacas — each with their own story. The most interesting calls always start the same way:

"We've got a wild alpaca. It spits, it runs at us, it can't be shorn, and we can't manage it."

The first problem is right there — they said alpaca, not alpacas.

Having one alpaca by itself instantly cuts its life expectancy in half. You've asked it to work twenty-four hours a day, seven days a week — no rest, no backup, no herd, no one who speaks its language.

Tell me — could you do that?

Now tell me your attitude wouldn't be atrocious too if you were highly strung, stressed, sleep-deprived, and completely alone.

We catch these animals, bring them home, introduce them to a herd, shear them, trim their toenails, give them a vitamin boost, feed them, talk to them — and within forty-eight hours, I've got them eating out of my hand.

And you told me that animal was wild and uncontrollable.

One of my favourite things is introducing new alpacas into the herd. It's like arriving at a party for the first time — a flurry of introductions, excited chatter, a bit of showing around.

The family and friend groups separate at first, but over a couple of days you see each alpaca mingling and making friends, until their little group is completely woven into the larger herd.

Even years later, alpacas still recognise the herd they came from. You'll see them standing or sitting together. And every now and then, we get to watch an alpaca mum reunite with her grown babies. It's a beautiful thing to witness.

Just like alpacas, humans are herd animals. We think we can do it alone — and maybe we can, for a while. But long term, we need the comfort, joy, support, and safety that being with our kind brings.

Turns out, we all settle better when we're not standing alone.

Life has a way of telling you what you need to hear. You just have to listen.

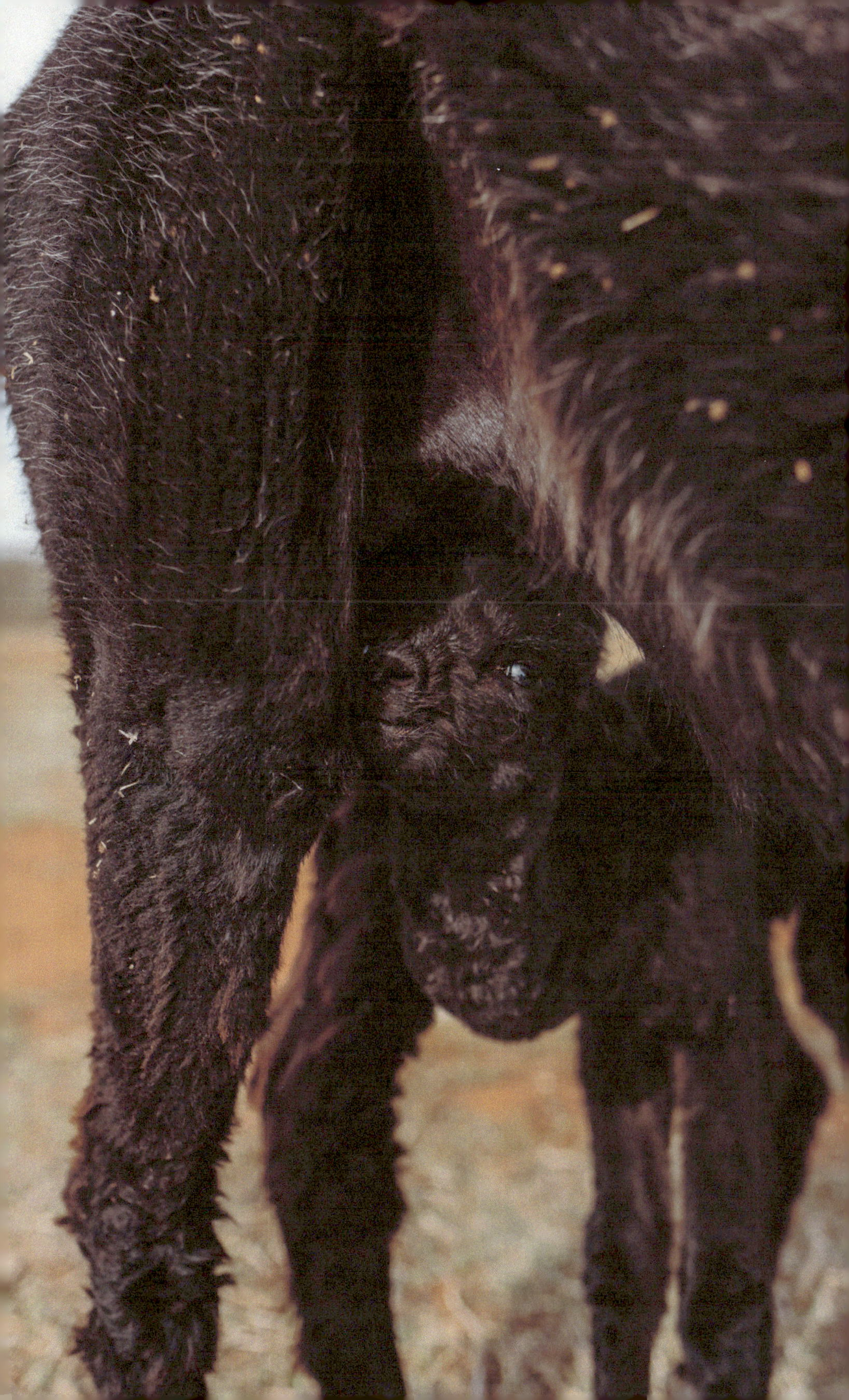

Love isn't about who's there when everything is easy. It's about who stays when it's not.

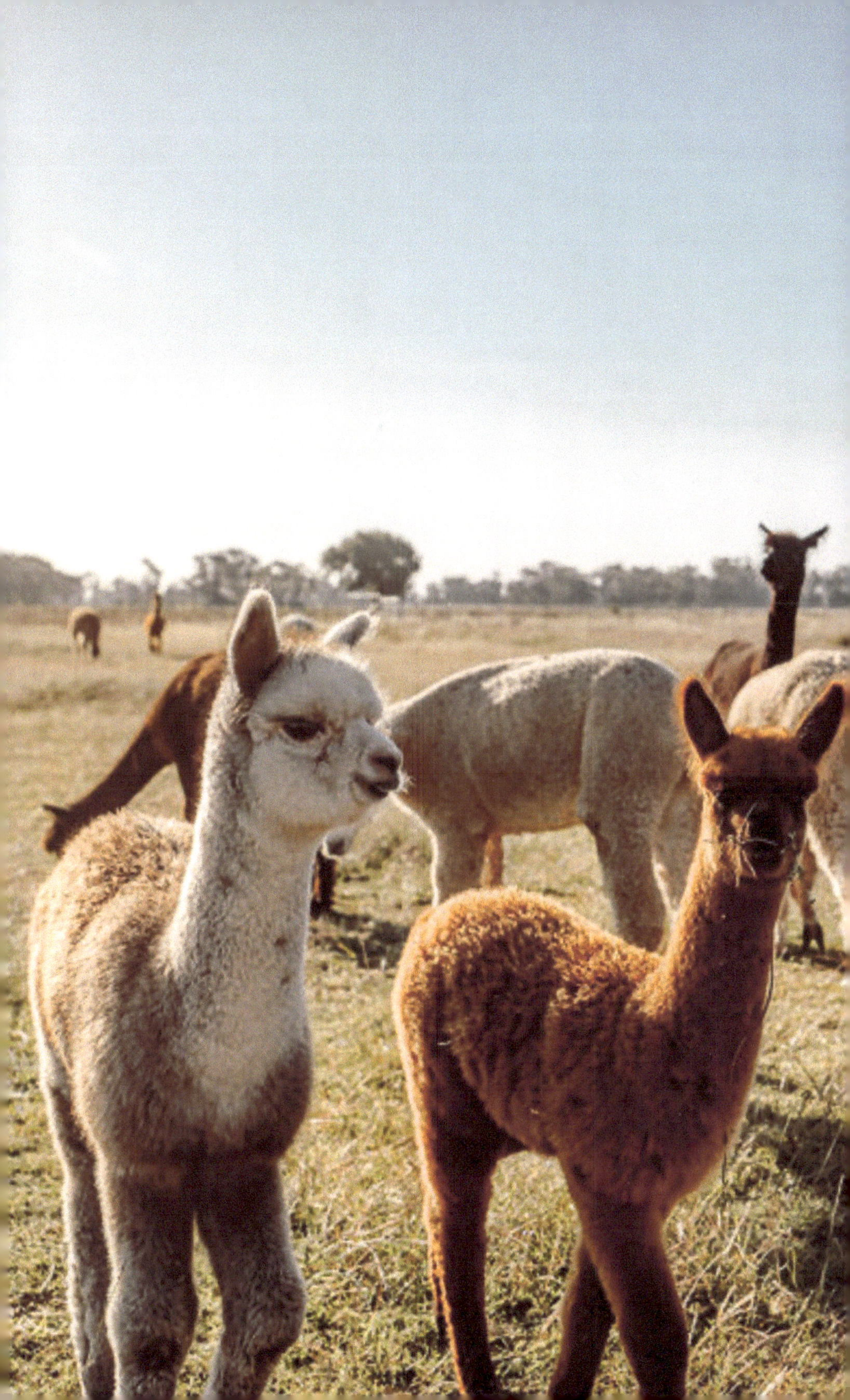

What you take in, you make room for. Choose wisely.

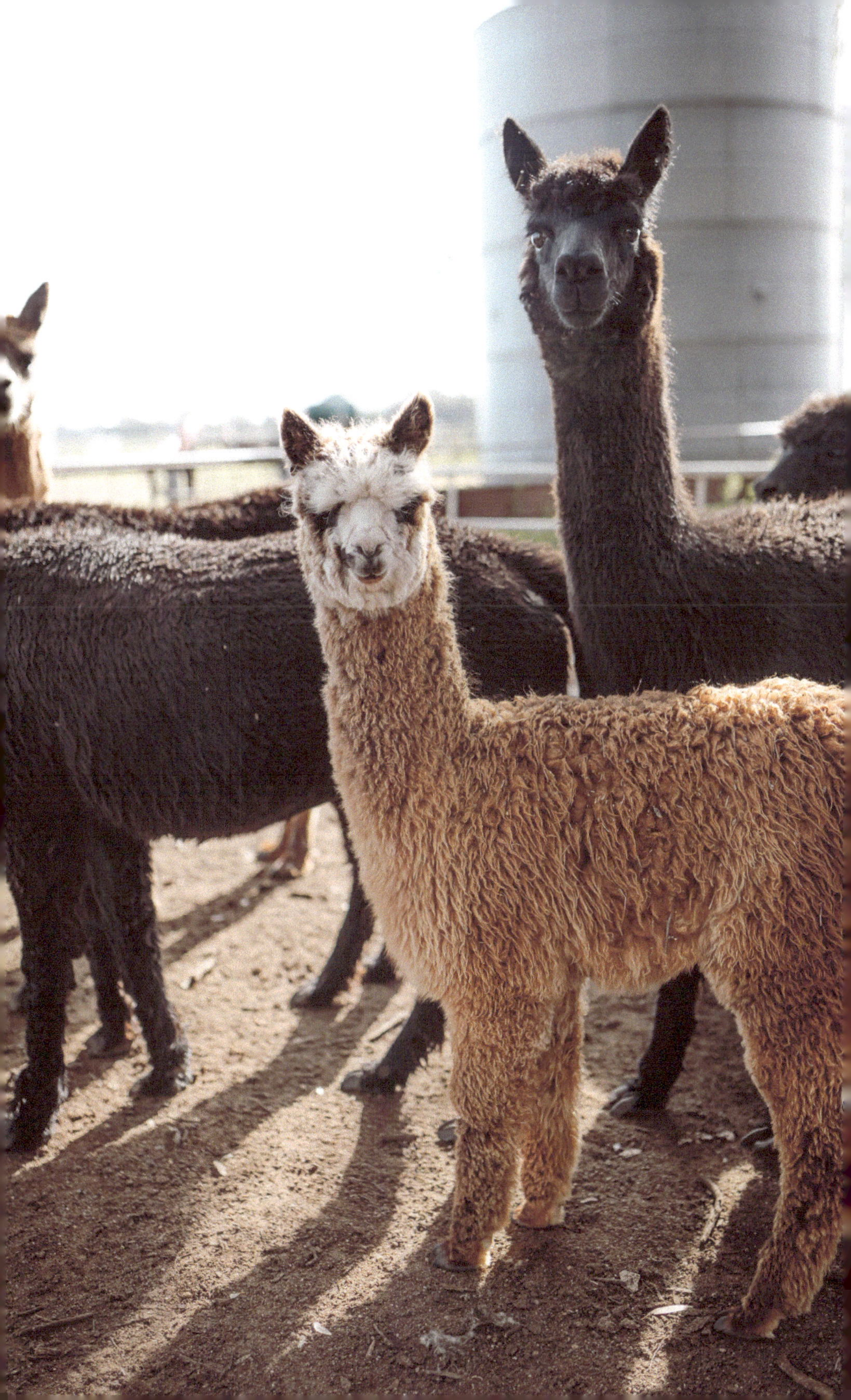

Good friends don't judge your bad hair days.
Great friends have them with you.

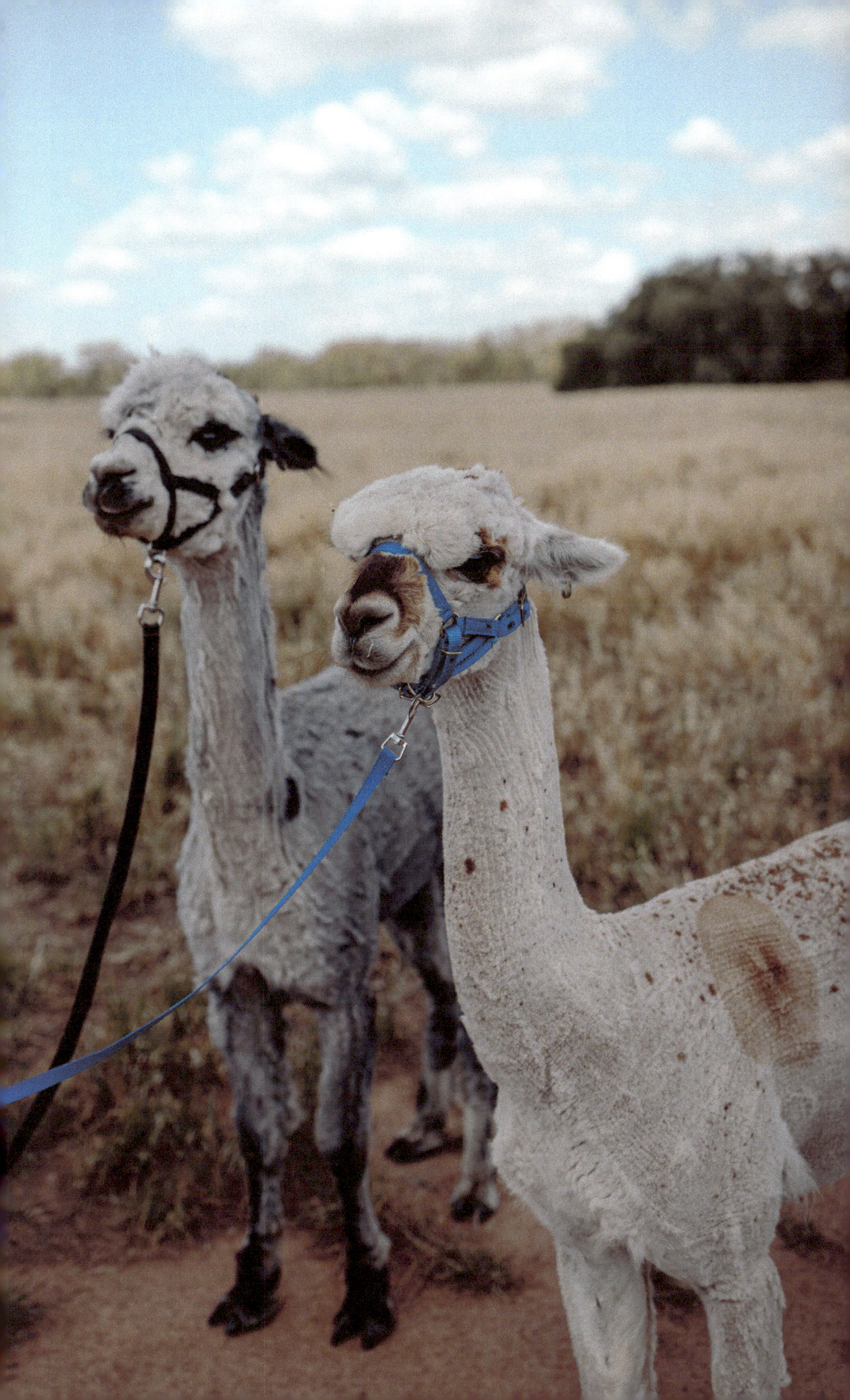

Stop waiting for a reason to celebrate. You are reason enough.

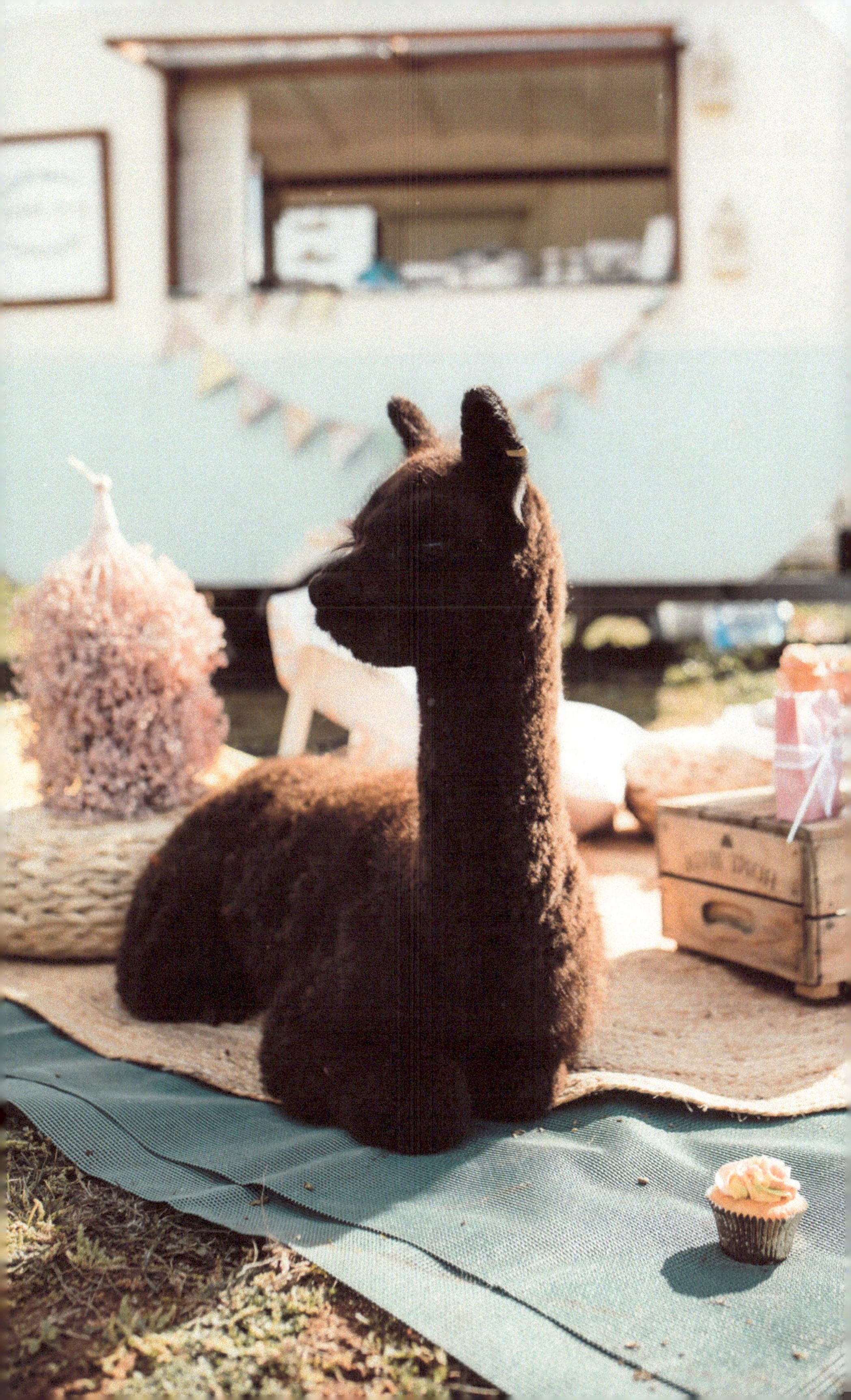

You get to choose who you surround yourself with. Pick the ones who make your world brighter.

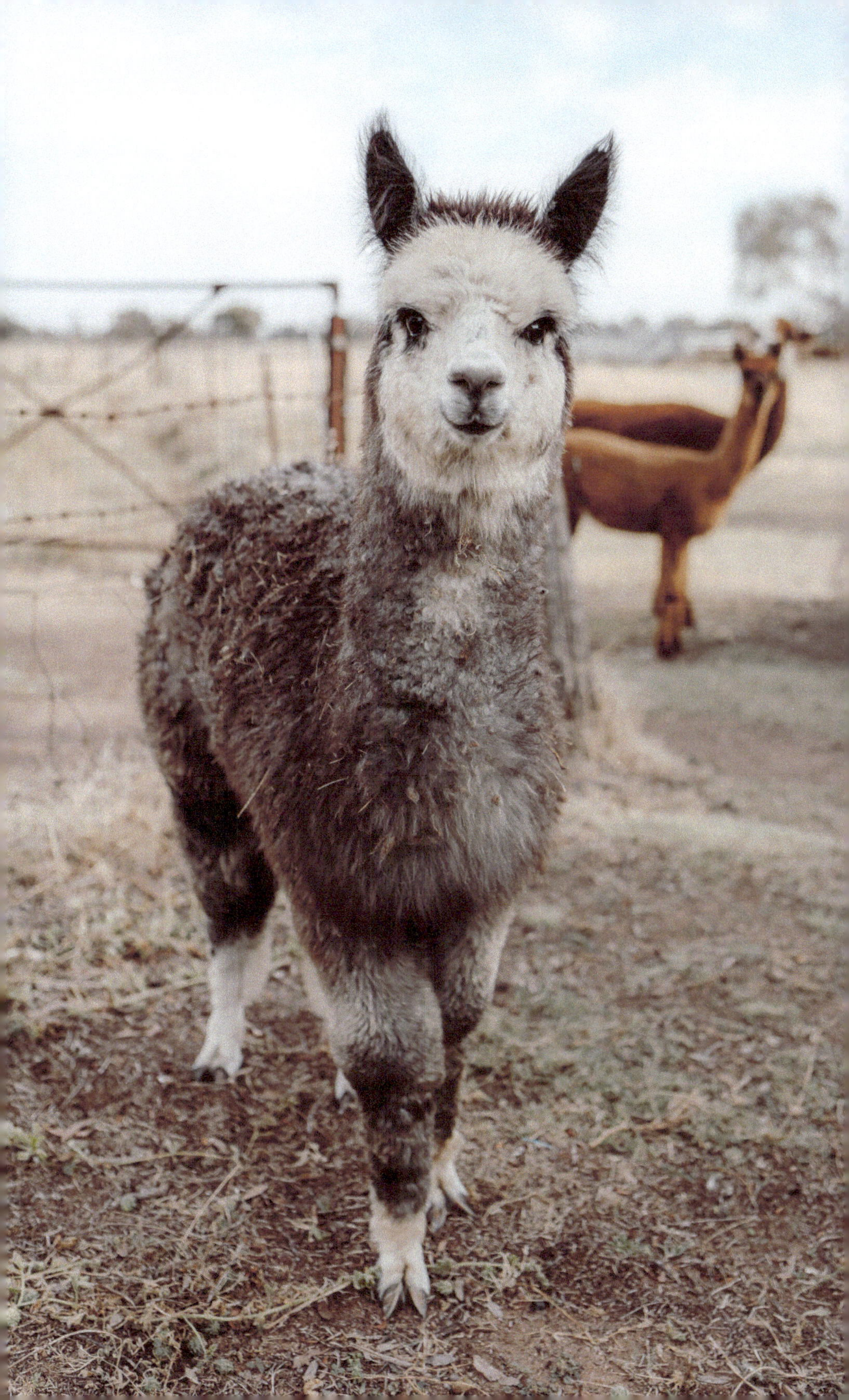

Kindness is never wasted. It always grows back.

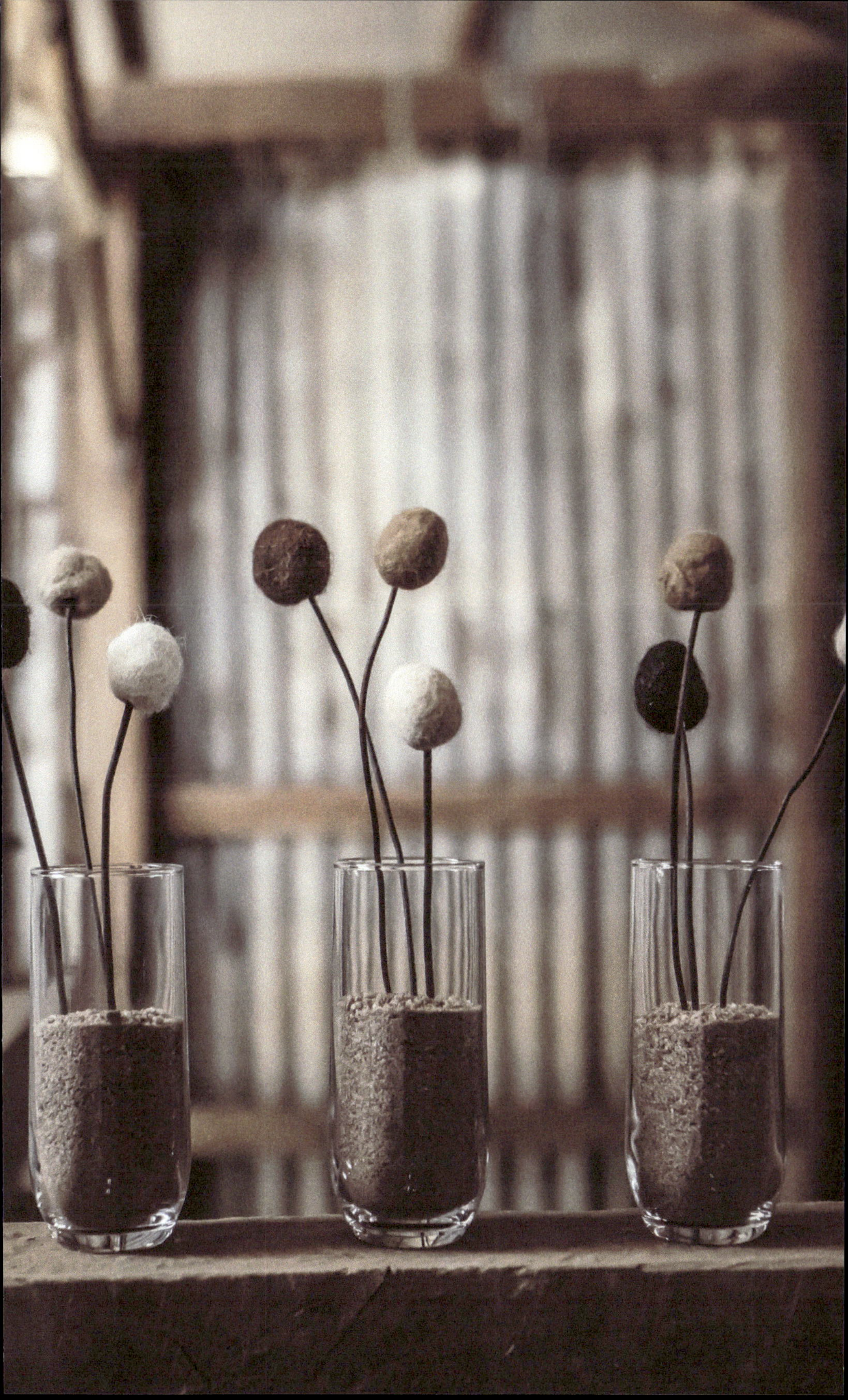

You don't always need words. Just company.

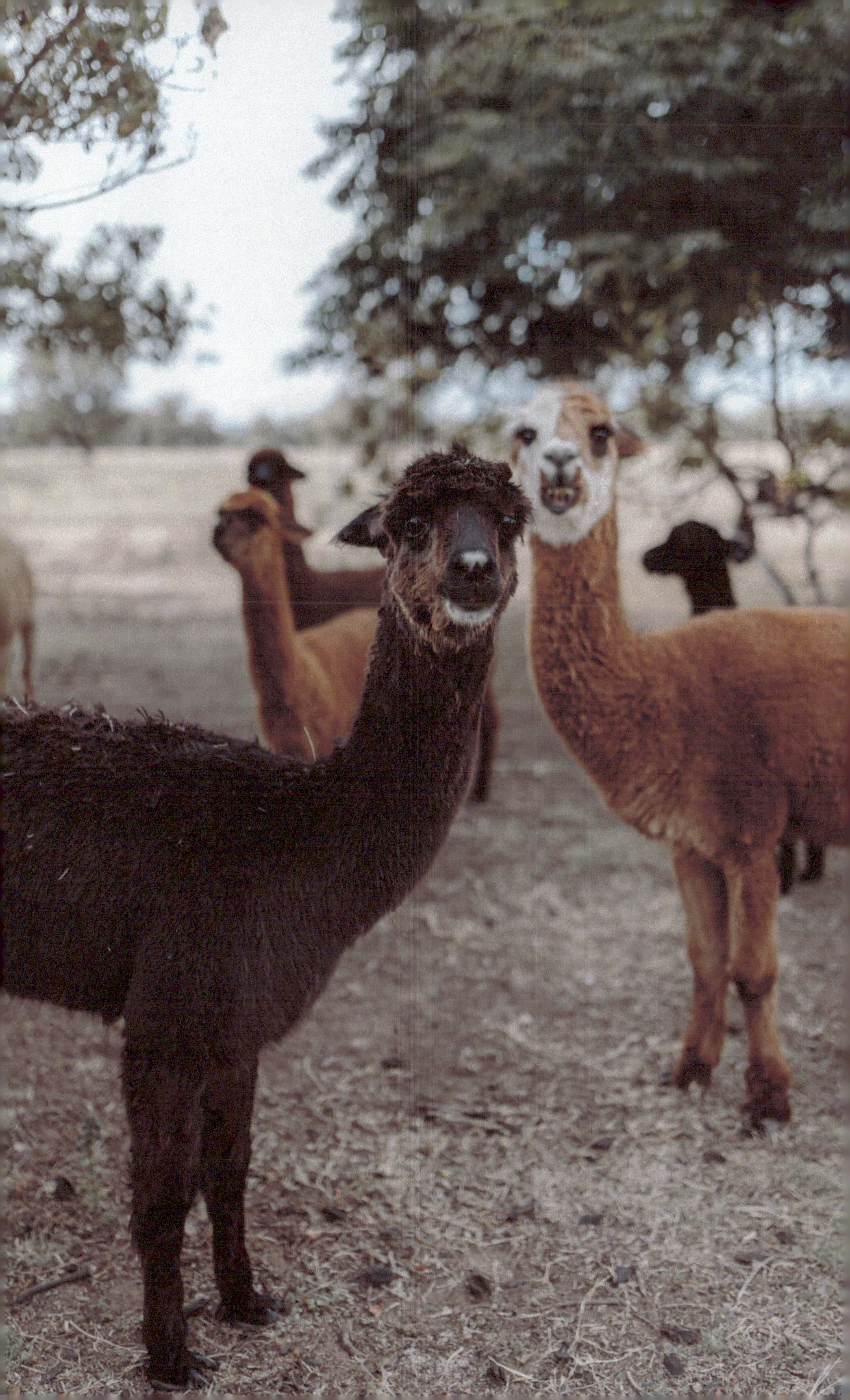

Gratitude turns an ordinary day into a story worth telling.

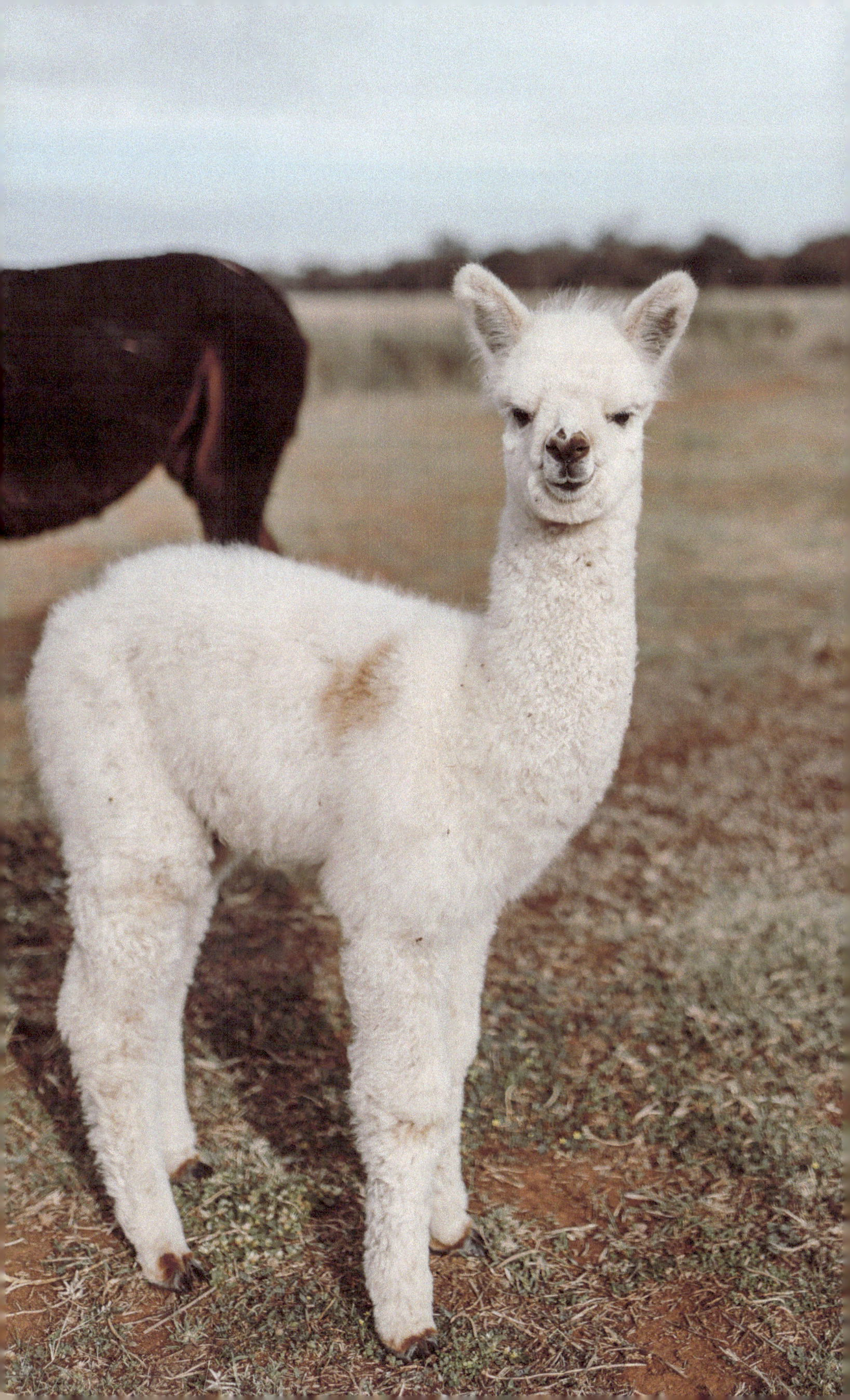

Don’t miss the magic just because it’s quiet.

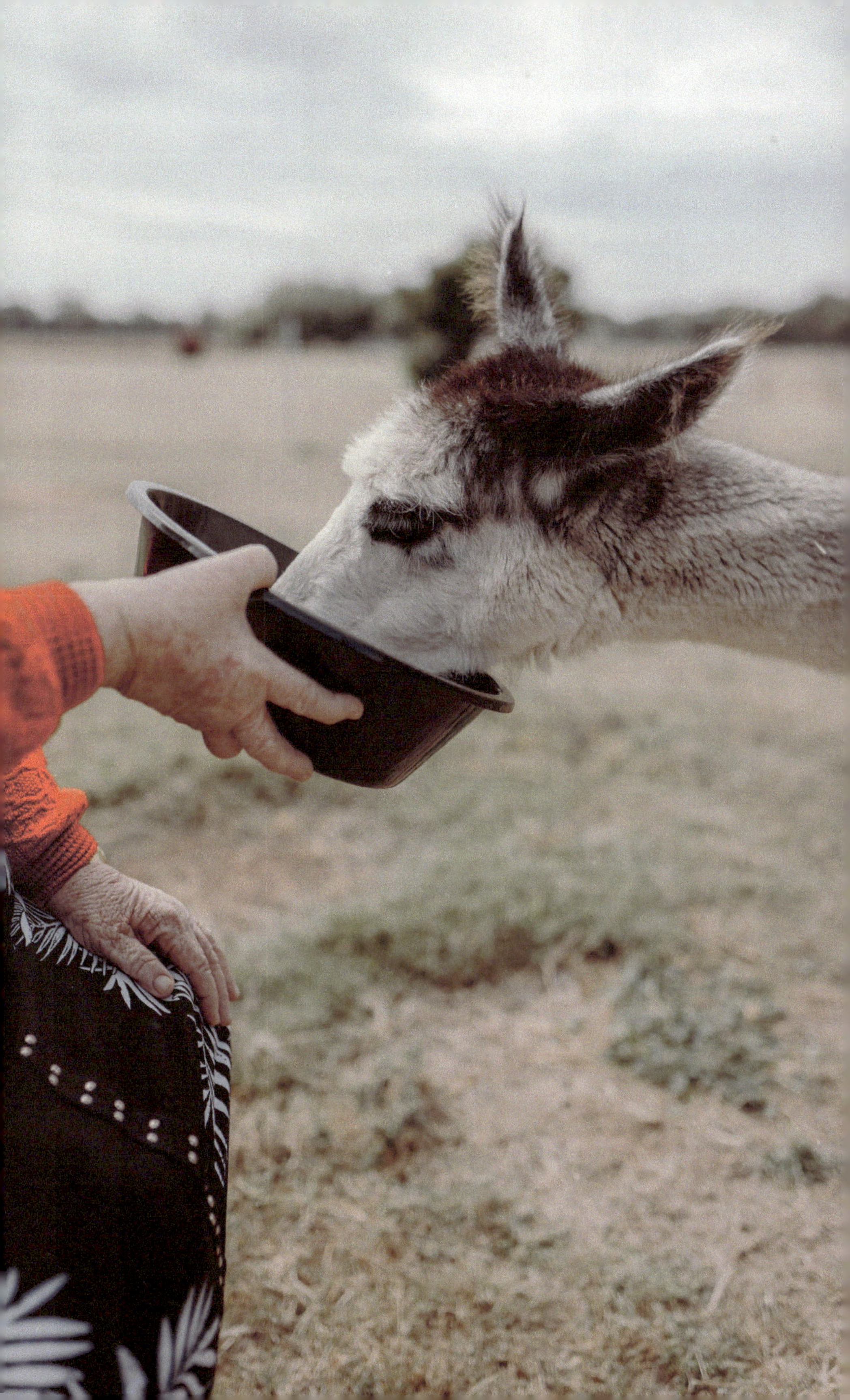

Perspective

Notes from the Paddock

Two questions I get asked all the time: “What’s an alpaca worth?” and “What’s a fleece worth?”

The answer is: it depends — on its value to your breeding program, on your farm, and whether you mean in the market generally... or to me specifically.

I’d been keeping an eye out for a grey alpaca (or two) to add to our herd. Ideally a male that could be a stud — to bring in new genetics, a few different colours, and more variety to our product range each year.

When a grey breeder announced they were retiring, I was thrilled. Their whole herd was going to auction. I read every description, made notes, set a budget, circled my favourites, and waited — impatiently — for auction day.

The auction began with that strange mix of excitement and mild panic that only happens when money and livestock are involved. A young grey male came up first — good breeding, nice fleece, solid build. I bid, won, and bought him for $500.

Another young male — similar colouring, similar breeding — sold for over six thousand dollars.

Welcome to the world of alpaca breeding. Where logic ends and personal preference takes the wheel.

When we went to pick up our new additions, curiosity got the better of me. I asked if I could see the one that sold for the crazy money. The owners pointed him out. I looked from him to mine and thought, Mine's prettier.

For someone else though, that boy had the right genetics, fleece density, micron, and colour profile — all the things that, for me, weren't as important.

If you visit the farm, I'll introduce you to Manrico. He's got the best 80s-rocker eye makeup you've ever seen, he's full of personality, and his fleece makes beautiful scarves and alpaca flowers every year. He'll eat out of your hand, and he's just had his first dates with a couple of girls. I reckon he's going to make very pretty babies.

But that's just my opinion.

Perspective is simply remembering what matters to you — and letting that be enough.

Sometimes, the best view is when you stop and see how far you've come.

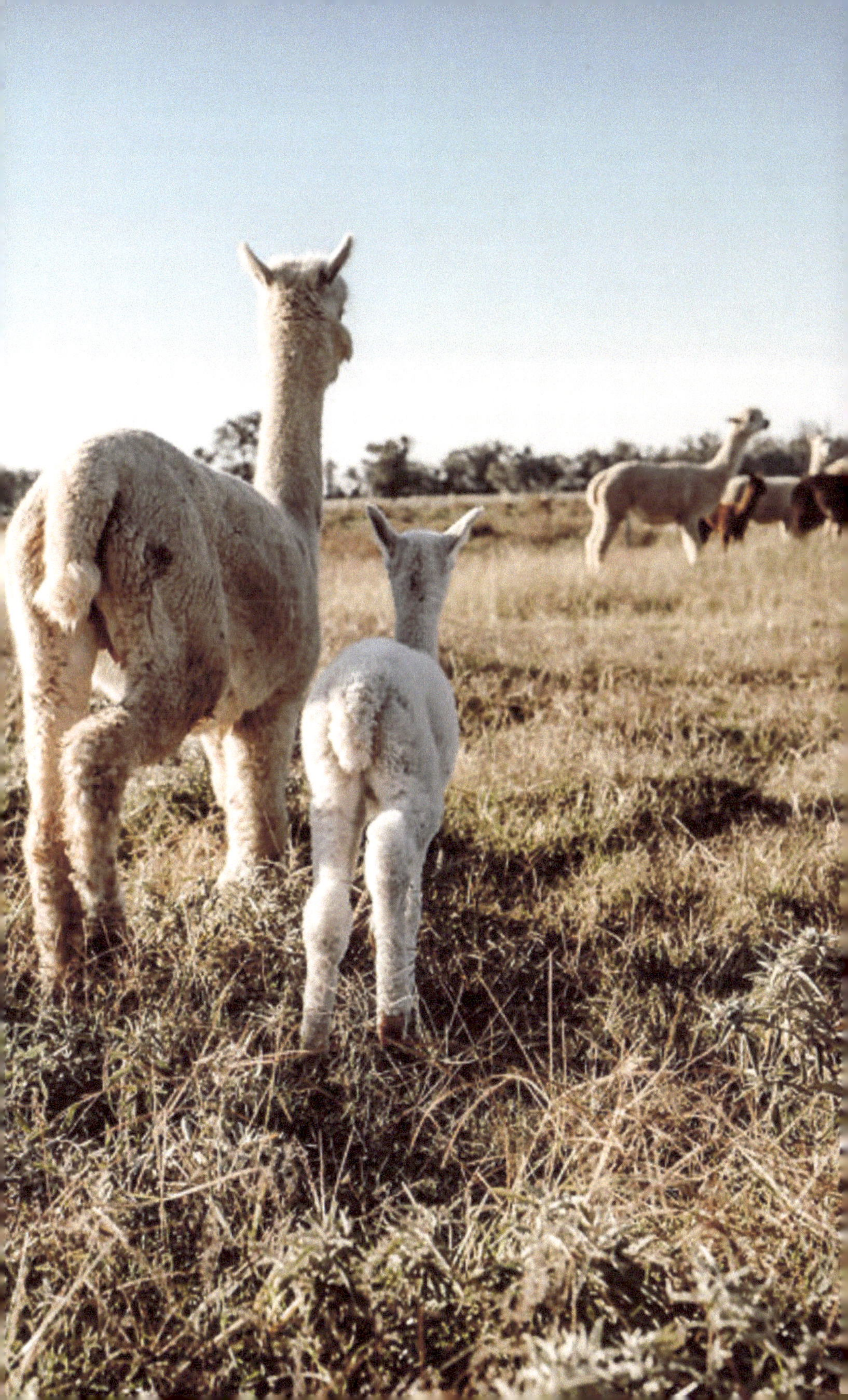

Stop. Look around. Somewhere in this moment, there is something good. Find it.

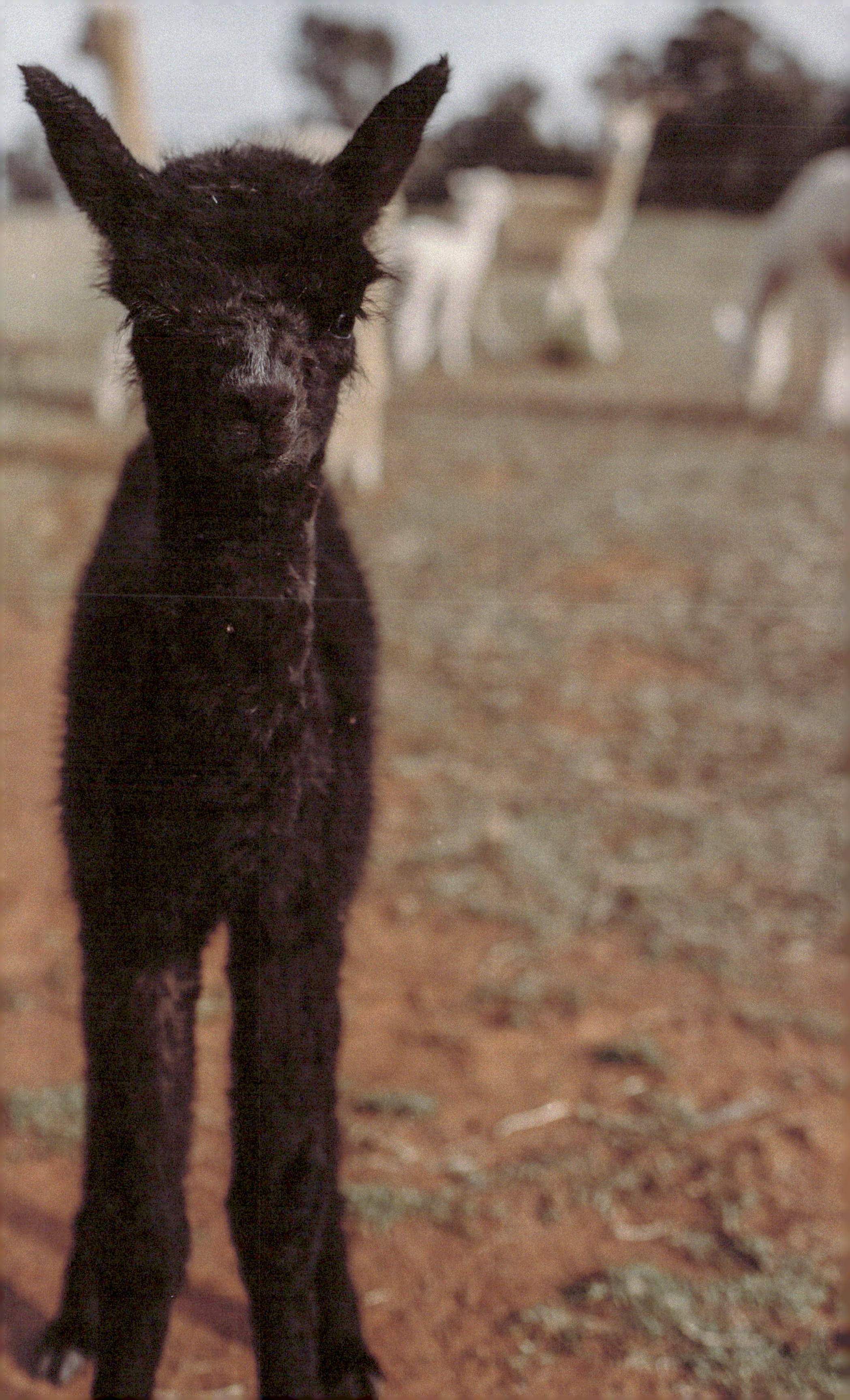

Letting go of the way things should be, makes room for the way things could be.

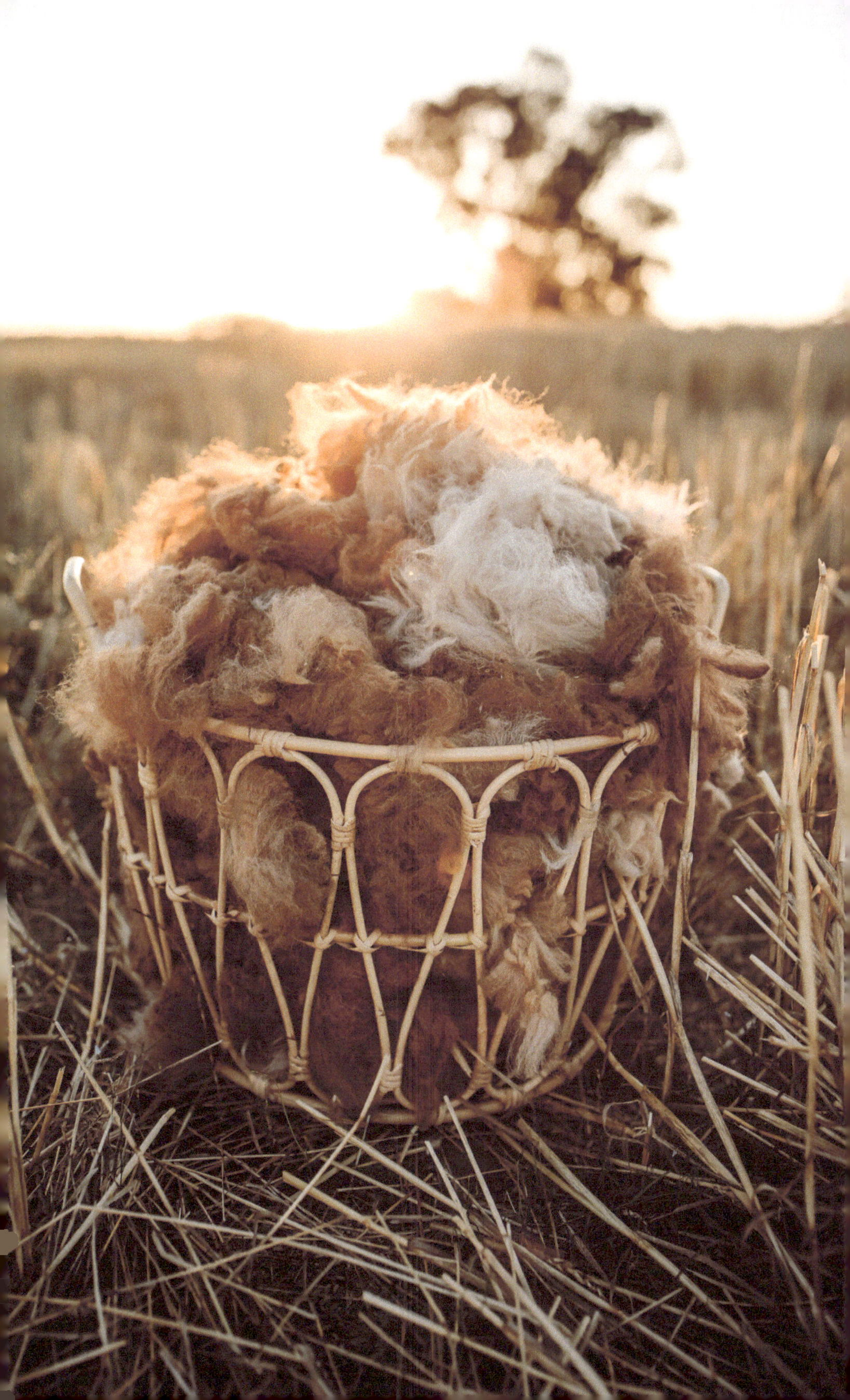

The beauty of life isn't in what you see, but how you choose to see it.

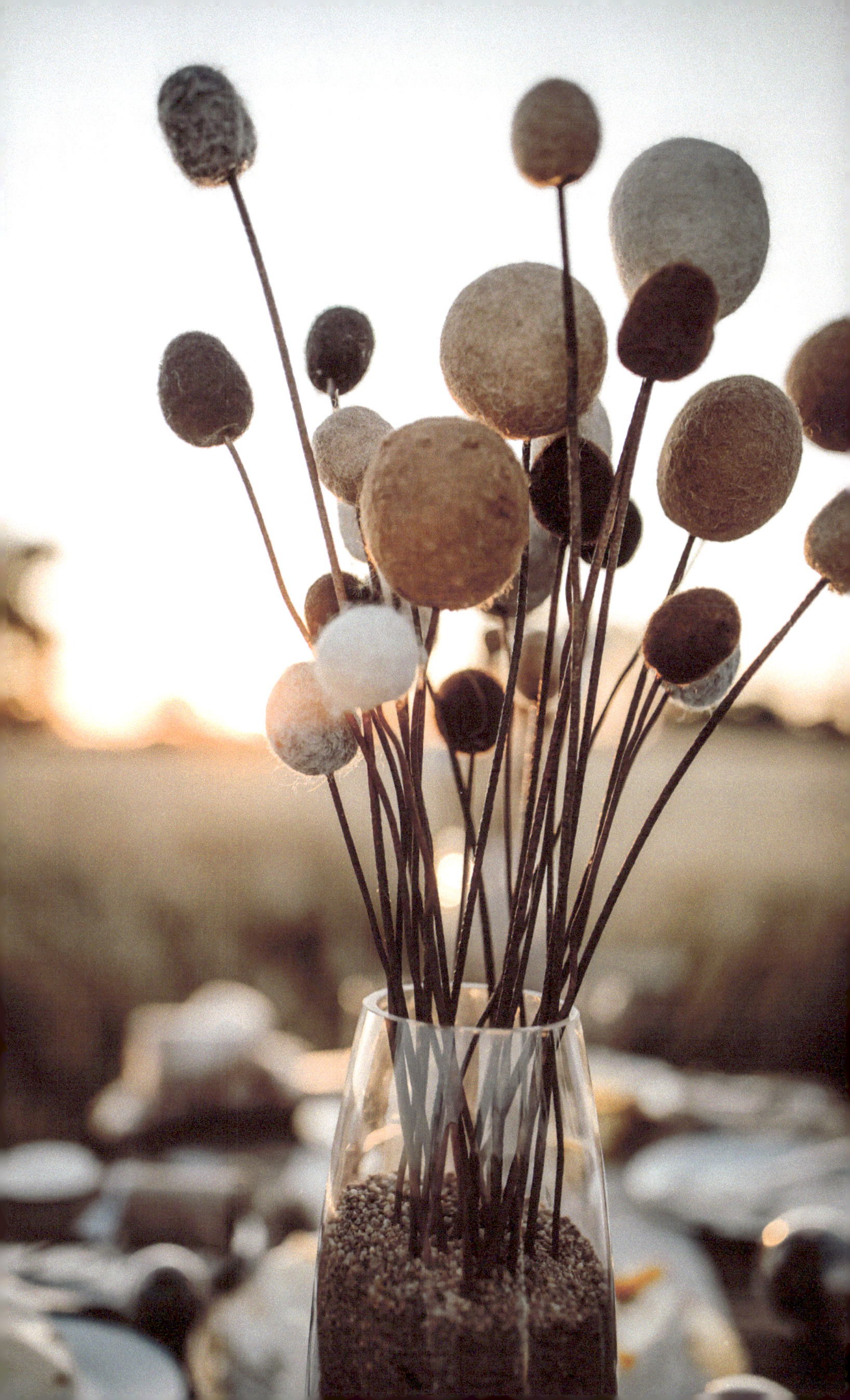

It's not about having space. It's about making space.

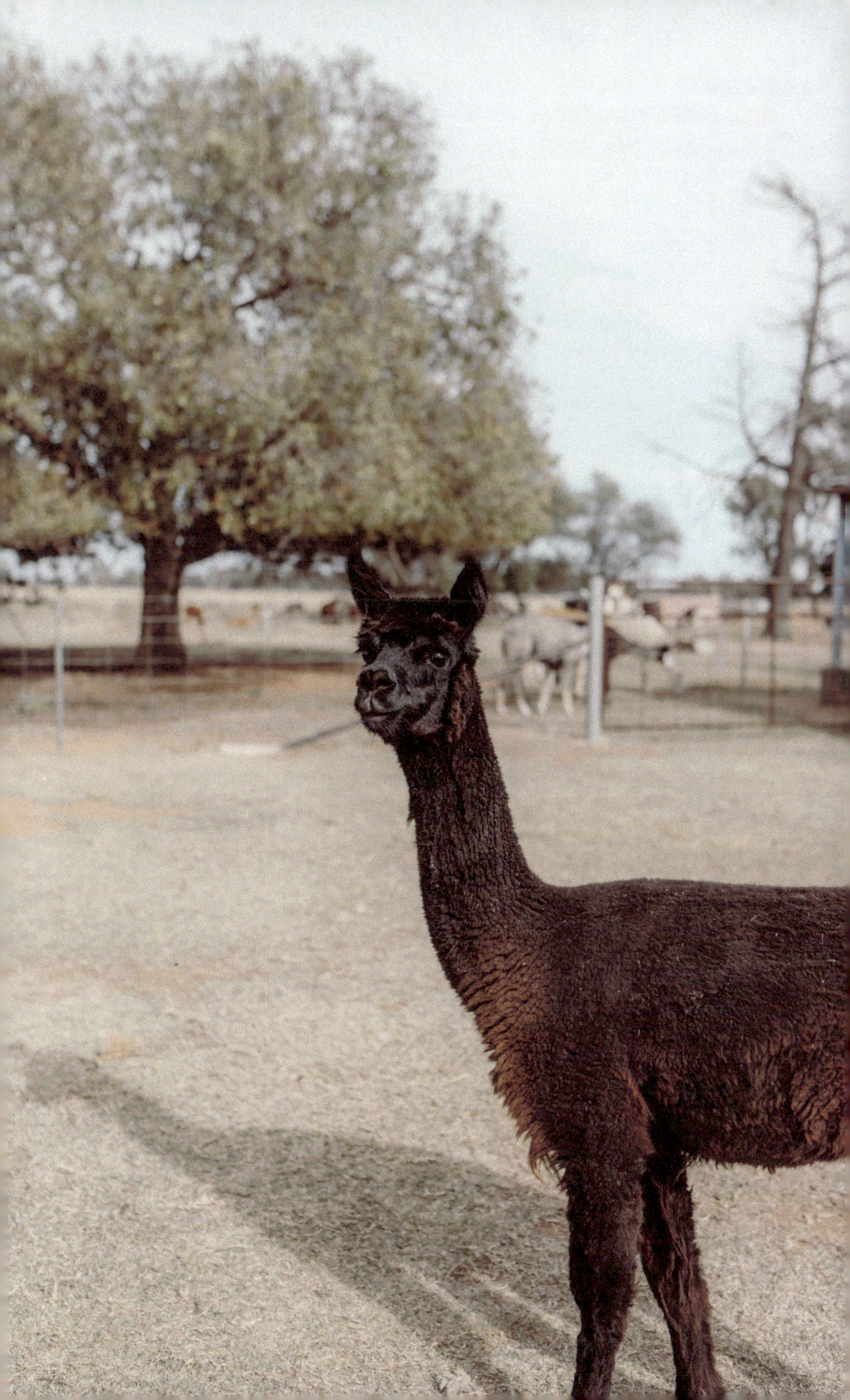

The sun still reaches you, even when you're standing in the shade.

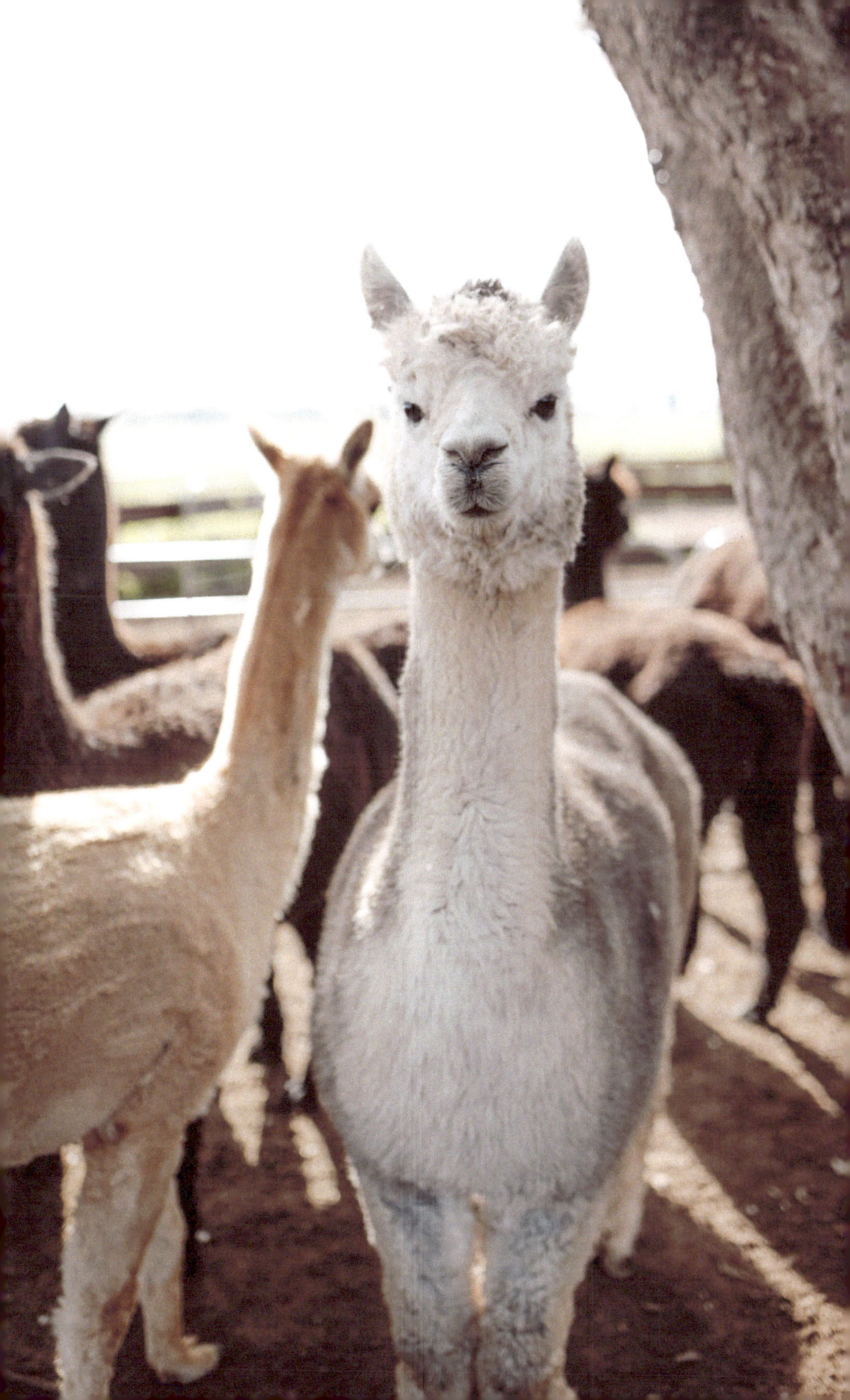

You don't have to have it all together. You just have to have it.

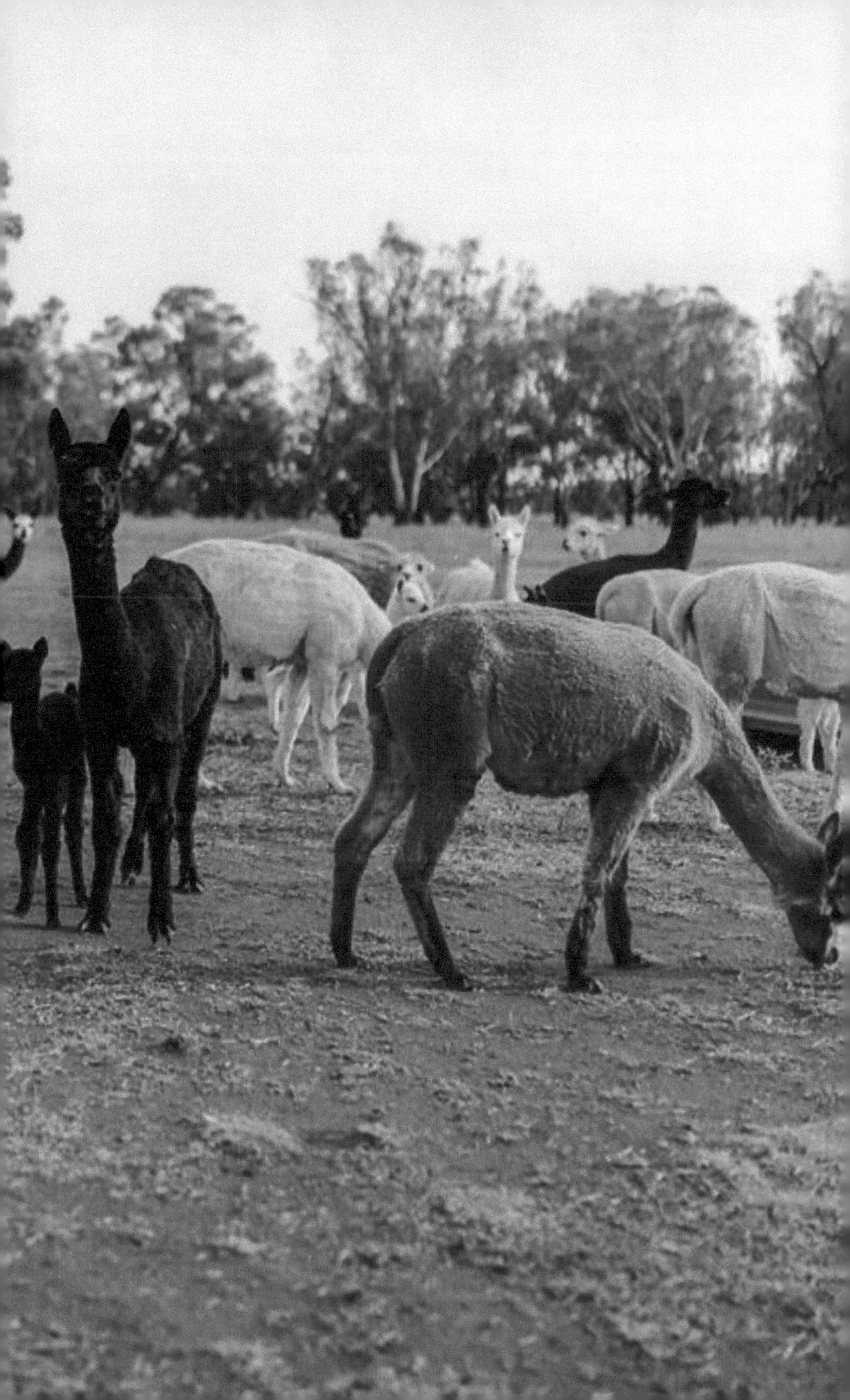

You are not here to blend in.

You are here to stand out.

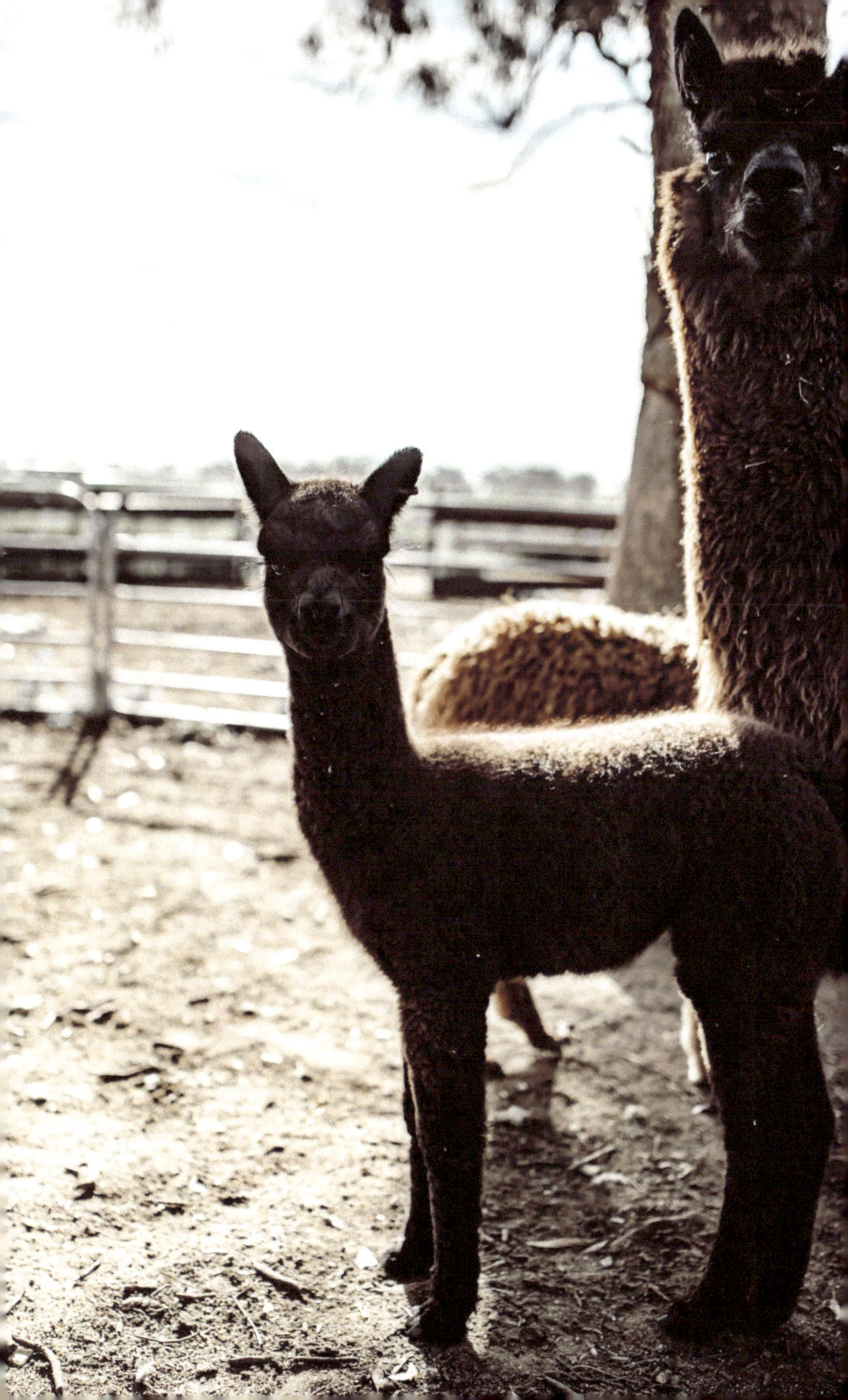

Find the extraordinary in the ordinary.

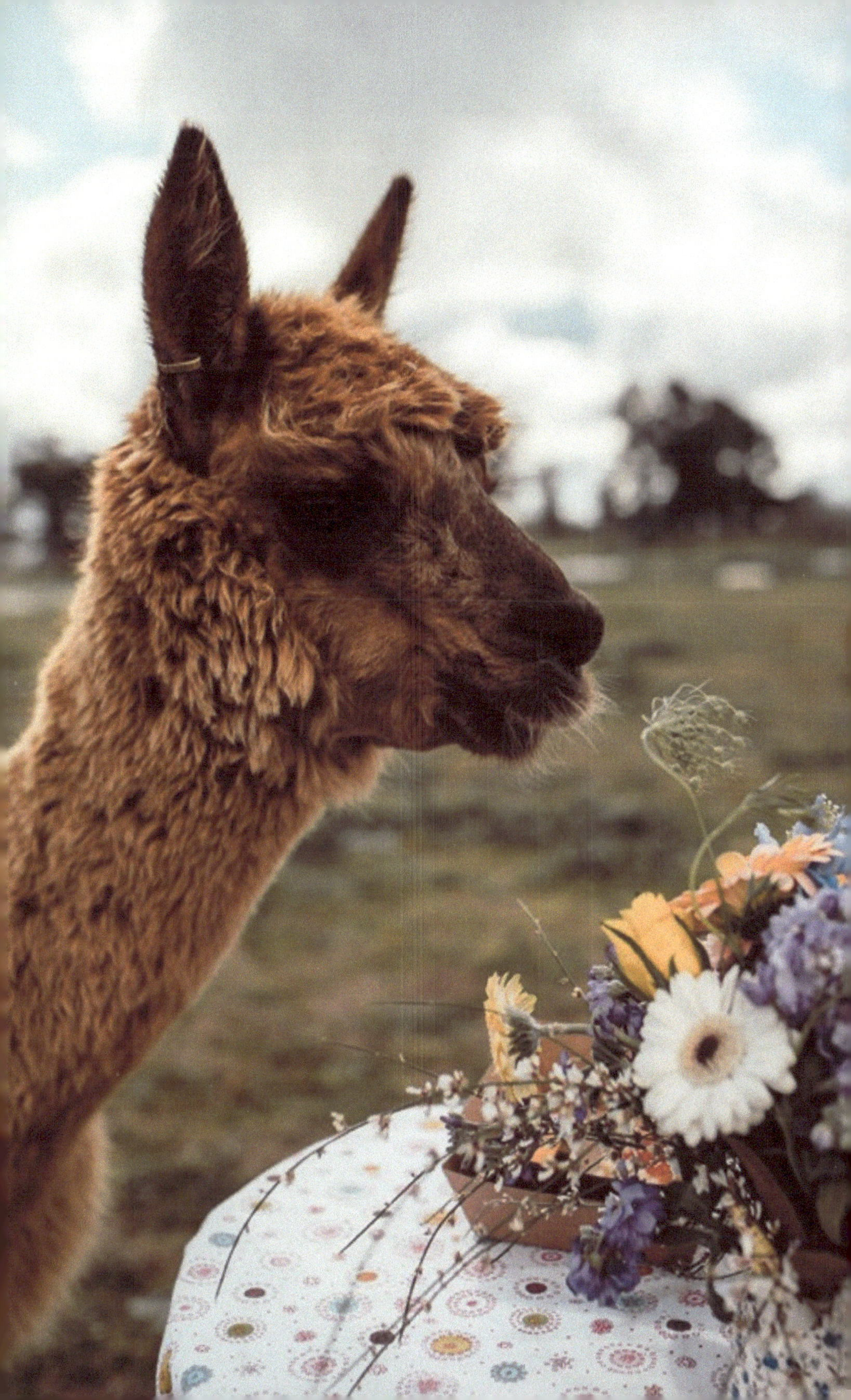

Pursue what lights you up, even if no one else understands it yet. They'll catch up.

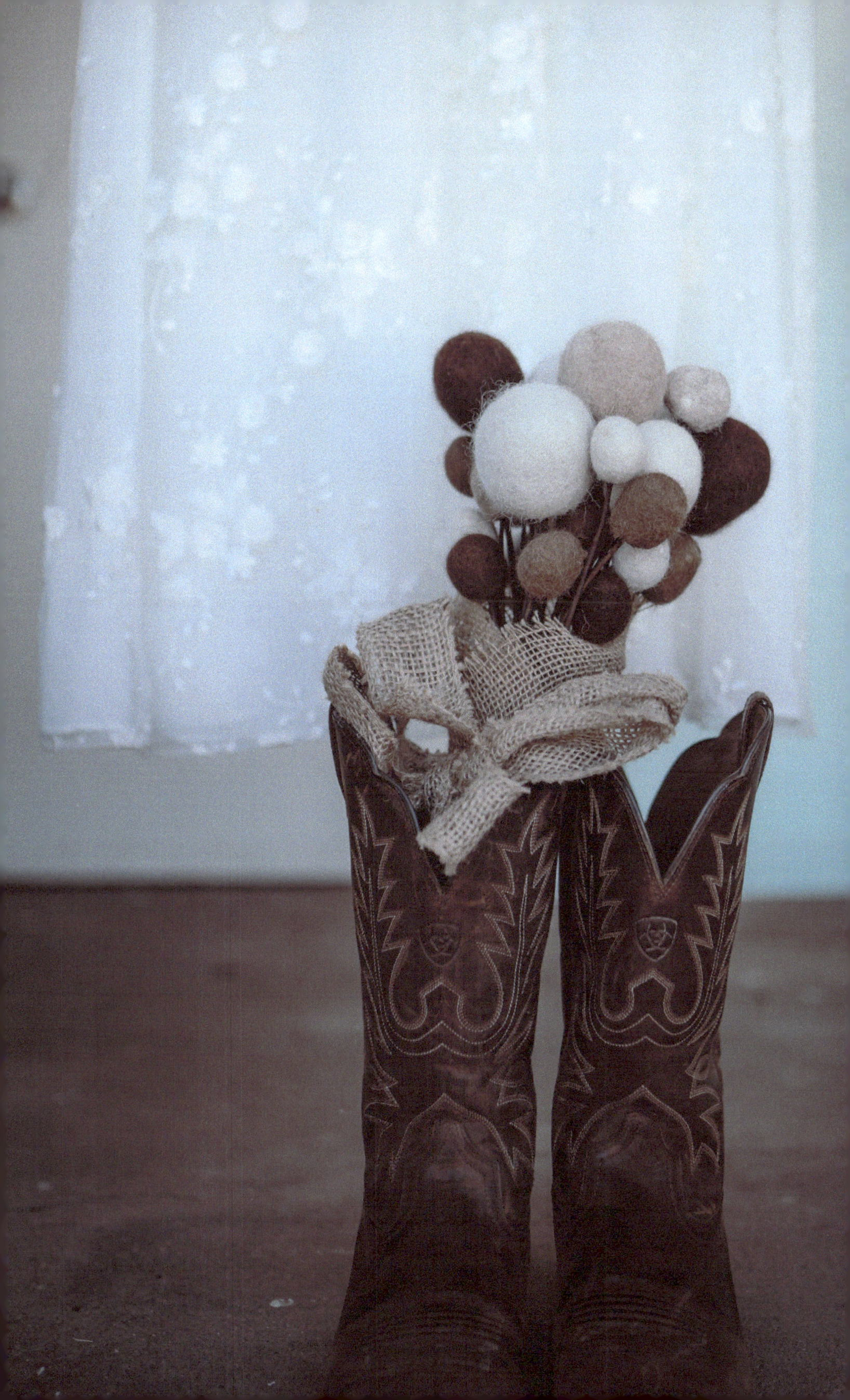

At the Fence

Between the parts

Part One was the view — the faces, the colours, the small moments that hold more meaning than they let on.

Those pages were the first lessons: the quiet ones, the ones that arrive without explanation — the kind you feel before you fully understand them.

Part Two is where those lessons explain themselves a bit more — or as much as they ever do.

Nothing here is polished.

Nothing here is profound.

It's just the little truths that bubble up when you're feeding animals, fixing things that broke themselves, juggling a dozen jobs, and laughing at the fact that none of this is going how you planned.

If Part One helped you step into and see the paddock,

Part Two will help you hear it — and maybe recognise something familiar in its voice.

Part 2

Insights

Daisee Press

Joy

Noticing the good in the everyday

Life on the land can be tough. I mean, really tough. It's heartbreaking. It's messy. It's unpredictable, and it's not for the fainthearted. I used to think anyone could do what we do, but the older I get, the more I realise this life is maybe not everyone's cup of tea.

We are constantly at the mercy of things far beyond our control: the weather, stock prices, feed costs... did I mention the weather?

During a particularly tough year, my mother started a little ritual with her grandbabies. As they said goodnight, she would ask, "What was the good thing today?"

At first it felt like such a small question for such big days.

But small questions can change everything.

There are three rules to this game:

1. It has to be from today. No digging up past glories.

2. It can't be food-related. ("Ice cream for dessert" doesn't count. That's cheating.)

3. You have to answer. Even on the days you don't feel like it. Especially on those days.

Occasionally a fourth rule sneaks in:

You can't copy someone else's answer. You have to find your own moment and share it.

Some all-time favourite "good things" from the kids include:

"Standing on the rails in the cattle yards and doing a bush wee on the cows!" This came from my nephew. Emma Daisy, my daughter, lost her nana at that point.

"That's not fair — how come boys can do that? Why can't girls?"

We laughed until our sides hurt. That was *our* good thing for the day.

"Playing in the puddles!" When drought is your normal setting, rain is a novelty, and we all go outside to take it in.

"Making quandong jam with Grandma." Some memories definitely taste as good as they feel.

"Going to bed, Nanna. My legs are just so tired." Mission accomplished.

The best phone calls are the ones that start with, "Ask me what my good thing is today," because you know whatever comes next will be worth celebrating.

This little game has given us moments of light in complete darkness.

I say game, but really, it's a deliberate, strategic, powerful way to shift perspective — to remind ourselves that joy is always there, even in the smallest of things. To take something good to bed with us, no matter how hard the day has been.

And now I'm sharing it with you.

(Remind me later to ask my mum if that's okay.)

Joy doesn't erase the hard days — it threads something good through them.

Joy doesn't need to be loud to be real.

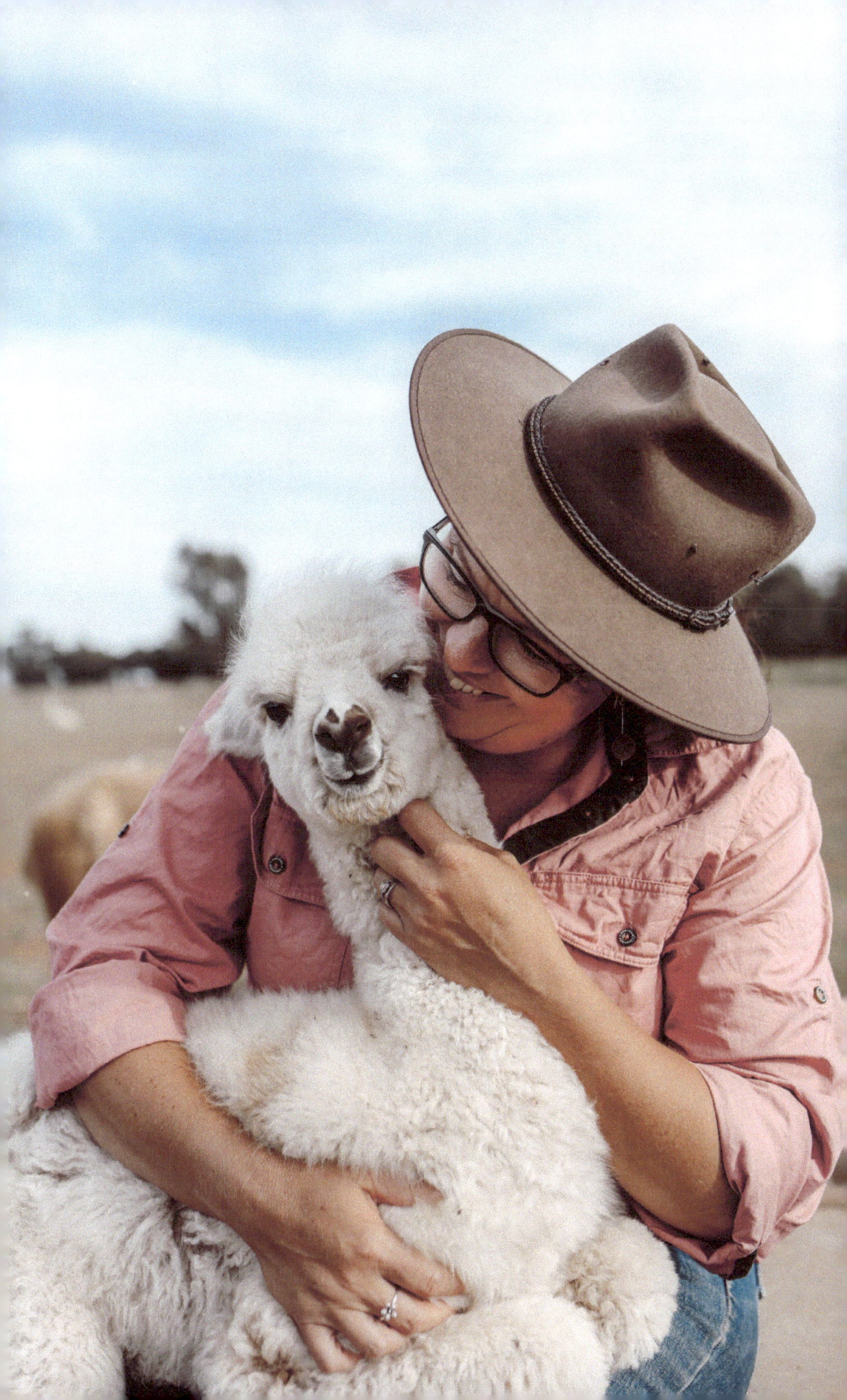

Opportunity

Opportunities are often disguised as challenges

Noticing an opportunity is one thing.

Recognising the opportunity hiding inside a challenge — and saying yes to it — is something else entirely.

Especially when it scares the absolute shit out of you.

Especially when you don't feel ready.

Especially when you have no idea how you'll pull it off.

If I've learned anything, it's this: the moment you find yourself thinking, *"I can't. I'm not ready. I don't know how,"* — that's the moment you need to say yes.

That's when doors open you didn't even know were there.

That's when you find a strength you didn't realise you had.

That's when you surprise yourself.

The magic, annoyingly, tends to happen the moment a deadline appears.

"Yes, we'd love to host your tour group on that day." (Meanwhile you're begging, borrowing, and rehoming enough tables and chairs to seat a football club.)

"Yes, I can absolutely make you a necklace out of handmade paper for your exhibition opening." (Going from idea to something real, tangible, and hopefully wearable.)

"Yes, I'd love to be part of Sustainable Couture." After making one necklace, you now need to design and create an entire collection... in two weeks.)

"Yes, I can put together a book reflecting on what I've learned from life on the land and the alpacas." (And here we are.)

"Yes, I can make you a bunch of Alpaca Flowers to order." (And then waking up to eighty orders and realising the universe forgot to include the part where you also have to make them.)

I've never said yes because I felt ready.

I've said yes because I knew that if I waited for the perfect moment, I'd still be waiting.

Every opportunity I've taken has started the same way:

panic,

a deep breath,

and then doing the next tiny thing in front of me.

Somehow, that's always been enough.

And here's the thing: once you say yes — and recover from the initial oh-no-what-have-I-done moment — help almost always shows up.

It might be something you learn.

It might be a person who appears at exactly the right time.

It might be someone already in your life who quietly has the skills (or willingness) you need.

The trick is saying yes to that help too, in whatever form it arrives.

Because opportunities rarely arrive gift wrapped.

They show up disguised as challenges, obstacles, and, most often, really hard work.

The most important thing you can do is simply say yes.

Because when you do, you set a deadline for yourself —

and that's when the magic happens.

The magic happens when you say yes, even before you know how to do it.

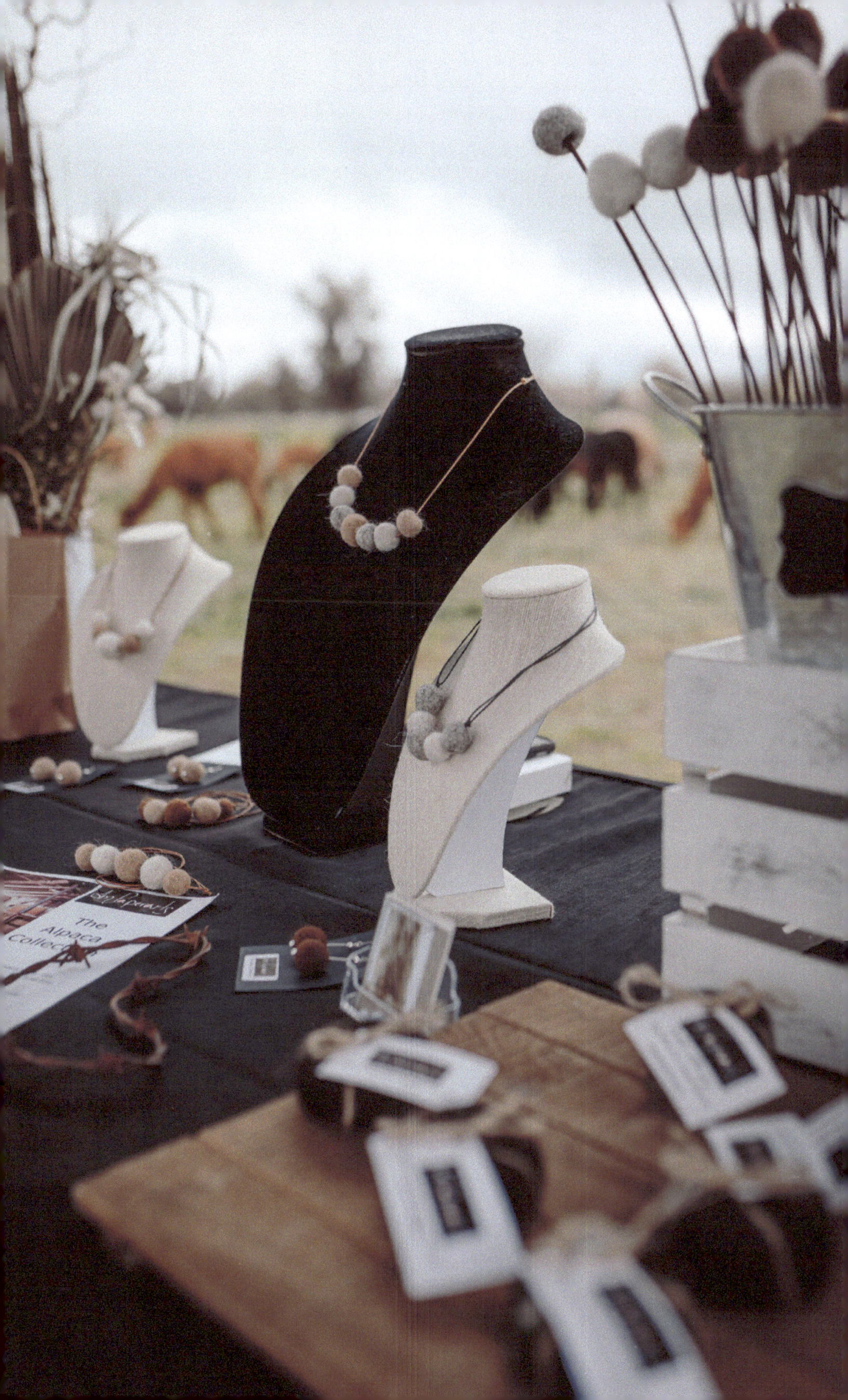
The
Alpaca

Resilience

Showing up anyway

Resilience isn't about never falling. Strength isn't about always being strong.

Sometimes it's pure perseverance — getting back up even when you're tired, even when it's messy, even when you have no idea how you'll fix the thing that just broke.

Animals and nature are the quiet masters of resilience, and if you're paying attention, they'll teach you everything you need to know.

Day three of a baby alpaca's life is my favourite example.

On days one and two, they wobble and stumble around, learning what to do with those ridiculously long legs. Slowly, they get them coordinated.

But on day three? On day three, they discover those legs can go fast. Very fast.

The problem? They still haven't worked out how to stop.

They tumble. They crash into aunties. They run into fences. They tangle themselves in a heap of enthusiasm and chaos.

But you know what? They get back up — on legs that still seem to have a mind of their own — and they go again. And again.

By the end of the day, they don't just know how to run... they've learned how to stop.

That's resilience — not perfection, but persistence.

Last year, I let some cherry tomatoes fall to the ground, hoping a few seeds might survive winter. Hundreds of seeds sat in that garden bed, and when the next season came? Not one single plant grew there.

Instead, I found one — on the other side of the fence.

No watering. No protection. At the complete mercy of the chooks.

That one rogue plant is nearly as tall as me and producing fruit like it's out to prove a point.

Life really does find a way.

Farming doesn't wait for you to feel like it.

The day you're exhausted, over it, and barely holding yourself together will be the day the cows knock over a trough and you end up knee-deep in mud, swearing at the universe.

The fences still need checking. The animals still need feeding. The plants still need watering. The eggs still need collecting.

Every. Single. Day.

No matter how you feel, or how cold, or how wet, or how hot, dry, or windy it is.

One foot still has to go in front of the other. No one else is going to do it for you.

And then there's the kind of resilience people mean when they say, "You're so strong — you'll be okay."

I hate that phrase with a passion.

Why? Because sometimes I don't want to be strong. I don't want to get up. I don't want to be okay — not yet.

Sometimes I want to be a sulky, messy human who cries and eats ice cream for dinner.

I might look strong on the outside, but the moment the gate shuts behind someone, the tears fall.

But that's resilience too — recognising the moment you need to crumble, letting yourself feel it, taking a breath, dusting yourself off, and getting on with the next thing.

Resilience isn't about being unbreakable.
It's about being breakable — and showing up anyway.

Falling's part of it. Getting up is the art.

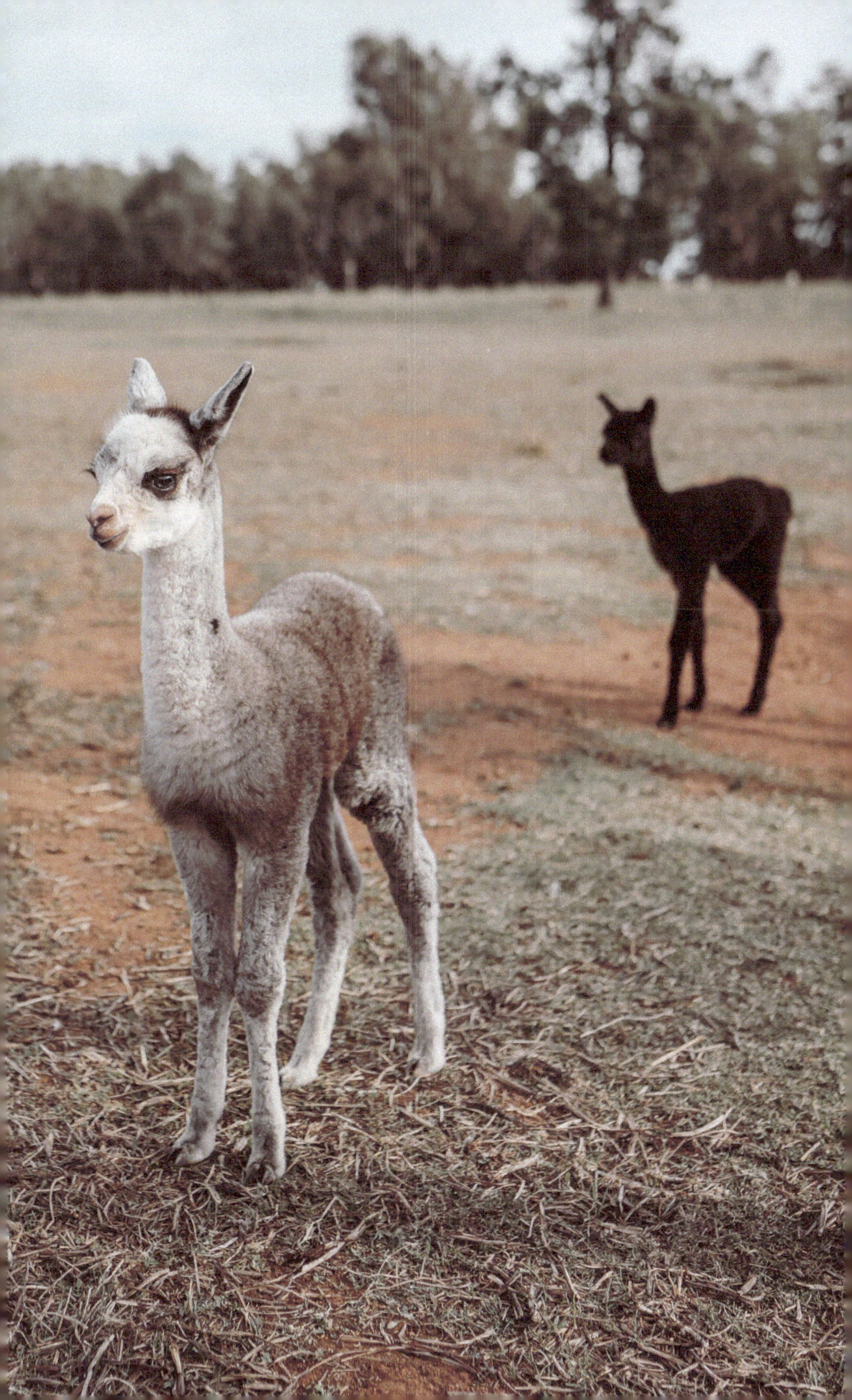

Hope

Believing in what you can't see yet

The most useful piece of clothing I own is a zip-up vest.

It has housed joeys, microbats, birds, and more tiny creatures than I'll ever admit publicly. It's warm, it's close to my heartbeat, and apparently it's the safest place in the world if you're a fragile little life trying to hang on.

And honestly, that's what hope looks like here — holding something small and vulnerable close, even when you know there's only a fifty–fifty chance.

Every time we bring an injured or orphaned baby inside — whether it's into the house or the unofficial hospital pen — the conversation is always the same.

No names on the first night. We've intervened, it's tiny, and there's only a fifty–fifty chance it'll see morning.

By day two, it has a name.

By day three, we're attached.

By day four, we're pretending we weren't attached on day one. (We absolutely were.)

Sometimes we get bold and say it's an eighty–twenty chance. But no matter the number, we are always — *always* — one hundred percent hoping for the best. Hoping it survives. Hoping it thrives. Hoping this tiny heartbeat becomes one more life out in the paddock.

Hope is stitched into everything we do here — quietly, stubbornly, often against our better judgement. It's in the feeding, the nursing, the checking, the repairing and the trying again tomorrow. Even when the odds aren't in our favour. Even when it hurts to keep caring.

Farmers complain constantly — about weather, prices, feed costs, and then the weather again — but underneath all that noise, we are not pessimists. We're realists with a reckless belief that tomorrow might still work out.

Hope isn't naive, and farmers aren't fools.

We know our animals, our land, our seasons, our limits. We know that sometimes, despite doing everything right, things still go wrong.

And yet — we carry on.

Not because we're sure of the outcome, but because the outcome matters.

Hope isn't pretending everything will be fine.

It's knowing things might not be fine — and choosing to care anyway.

Choosing to try anyway.

Choosing to believe in the thing you can't see yet.

Hope is the softest part of farming...

and somehow, it's also the strongest.

Hope isn't a feeling. It's a habit. It's the bit of you that keeps turning up.

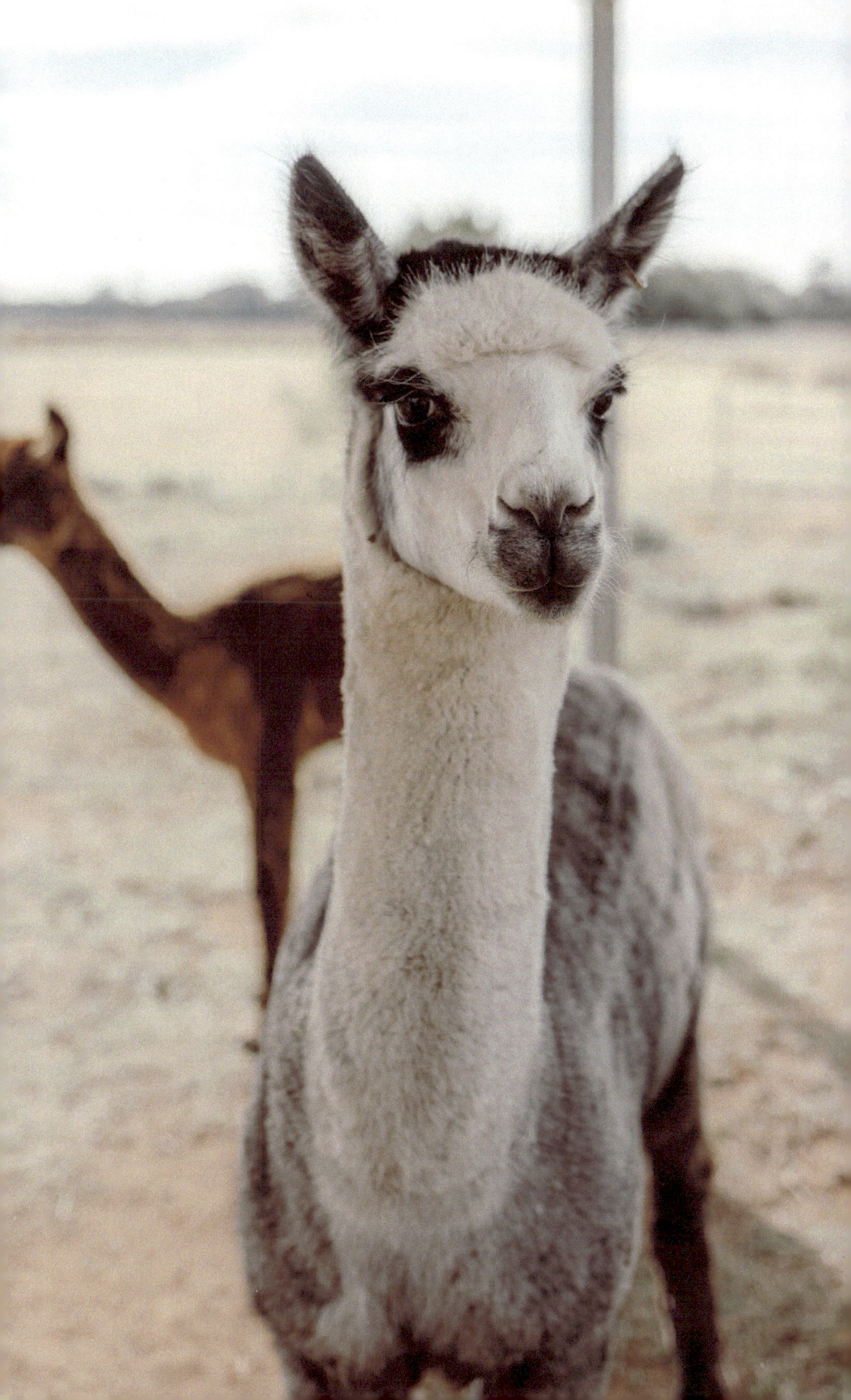

LOVE

Shown in the doing, not the saying

I wrote this down as a topic I wanted to include, and you'd think it would be the easy one.

It wasn't.

Love, as it turns out, is harder to define than drought, machinery failures, or a rogue animal with a personal vendetta.

I'm not an expert in love — far from it. But I've learned that it isn't one thing. It's felt differently, shown differently, understood differently. It's rarely the sweeping, cinematic version. Out here, it's quieter. Smaller. More honest.

On the farm, love is practical.

It's steady.

It's the kind of caring you can rely on, even if no one sees it happening.

It's knowing where you stand and standing there with certainty.

It's showing up, even when you're tired, or overwhelmed, or out of words.

It's doing the thing that needs doing, because something — or someone — depends on it.

Farmers face heartache more days than we'd ever confess to. Sometimes we have to switch emotion off entirely just to make the best decision for our animals, our land, our livelihoods, and ourselves. It's a learned skill — and like all learned skills, it doesn't always work when we need it to.

Some might say that dialling down or switching off the emotion would shrink your capacity to love. I've got two ex-husbands who'd probably agree.

But I also have a partner now who understands something important: his corner of my heart is his — but it's one of many corners — and that's completely okay.

Today, love looked like a water tank.

Trivial, maybe — unless you're a farmer.

A ten-minute freak windstorm left us with debris everywhere and one casualty: a second-hand tank that had arrived only the day before. And it wasn't just a tank. It was security — extra water for the troughs on forty-degree days, a step toward irrigating a paddock in drought, a tool to build a better environment for the animals who feed our family.

When the tank split open, it felt like something in me split too.

Sometimes that's what love is — pouring everything you have into something you know could be undone in a single storm you never saw coming.

And sometimes, love is holding an animal in the sun for its final moments because you know it's the kindest thing you can do. Sometimes the hardest decision is also the right one. It could be an animal you've had for five minutes or one you've raised longer than your own children — it never gets easier.

But love isn't always heartbreak. Most days, it's quieter than that.

It's heading back out at midnight because something in your gut says, "just check."

It's doing the unglamorous things, the uncelebrated things, the things no one claps for.

It's making a cuppa, sharing the last biscuit, and laughing at things that would make other people cry.

It's the small jobs done side by side, the quiet reassurances, the way life feels easier — or at least possible — when someone is standing in it with you.

Love isn't about perfection.

It's about presence.

Real love lives in the small jobs, done side by side, and in the laughter that comes after the hard bit's done.

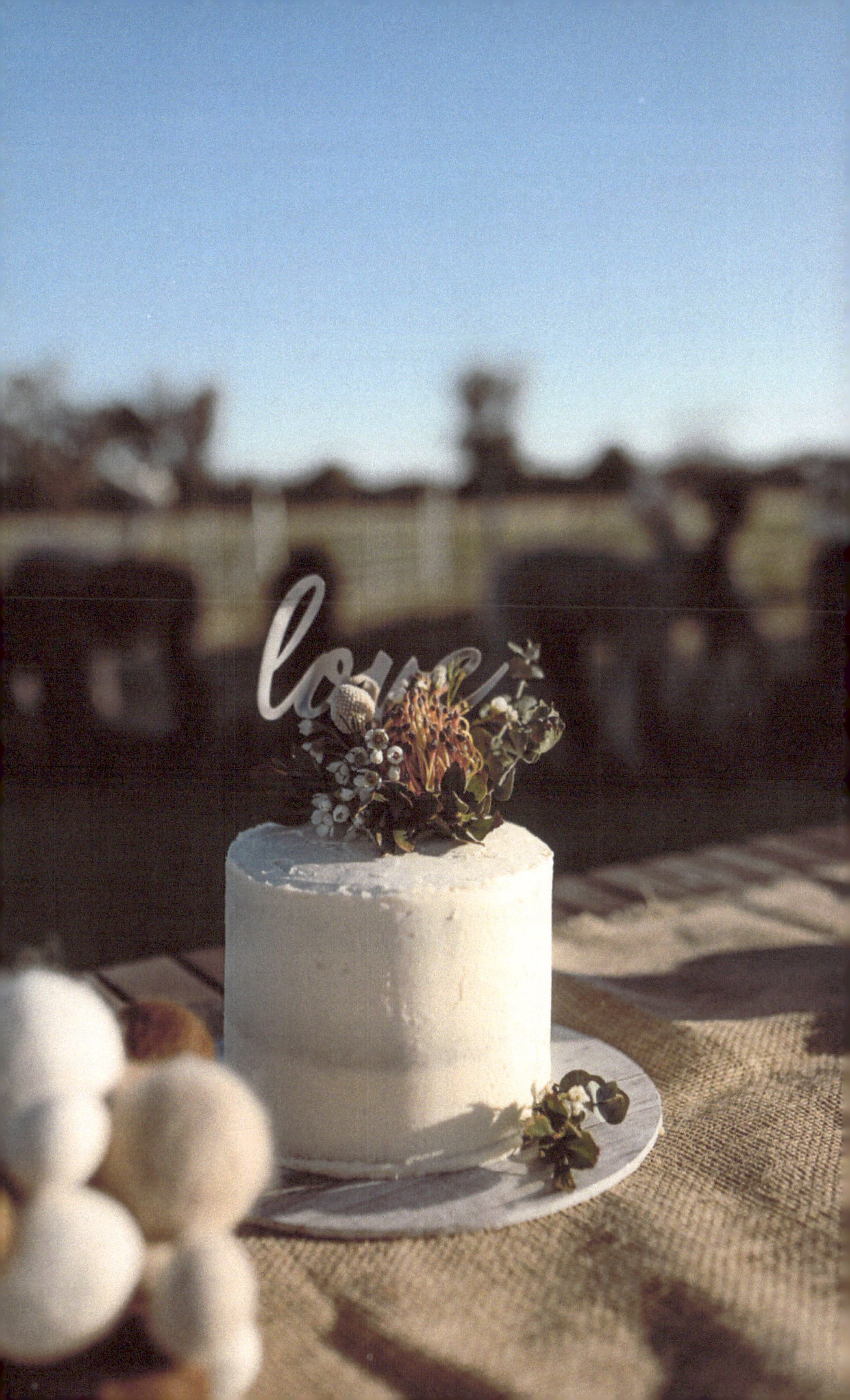
love

Reality

The bit that never makes the brochure

Life on the farm isn't all green paddocks, warm sunsets, and photogenic alpacas staring soulfully into the camera. Most days it's dust in places you'd rather not talk about, a to-do list that multiplies overnight, and animals who wake up determined to test your patience, your reflexes, and your vocabulary.

Reality here is equal parts beautiful, ridiculous and unbelievable.

It's catching an alpaca mid-birth while you're wearing town clothes and running late.

It's fixing a busted pipe at 11 p.m. with a headlamp that keeps sliding off your forehead.

It's being halfway through rolling out hay before realising the string is wrapped around your ankle.

It's getting a call from the neighbours to say your cows have escaped — again — and eaten everything in their veggie patch.

It's trying to look professional on a Zoom meeting when the goat lets himself in the back door.

It's also the reason I will never — ever — be able to own white clothing again.

People imagine farm life is peaceful and therapeutic, or that's what the brochure says, and sometimes it is. But most days happen in the in-between: the chaos between the calm, the "you've got to be kidding me" moments, the days when something breaks simply because you looked at it wrong.

Animals don't read your schedule.

Weather doesn't care about your plans.

Equipment breaks precisely when you need it.

Reality has a way of humbling you — reminding you who's actually running the show (hint: it's not you).

And yet, it's in those messy, unfiltered moments that the real story lives.

It's the laughter that bubbles up once the disaster has passed.

It's the pride in fixing something you had no business knowing how to fix.

It's the quiet triumph of making it through a day that tried very hard to break you.

It's standing in the last light of the evening, watching the shadows grow long and the animals settle down for the night, knowing somehow you made it.

And it's the certainty that tomorrow — no matter what happens — you'll show up again.

Farm life is imperfect, unpredictable, occasionally absurd... and entirely real.

Reality isn't the part we should filter out — it's the part that makes everything else mean something.

If you can't laugh at it now, give it time. One day, this will be a great story.

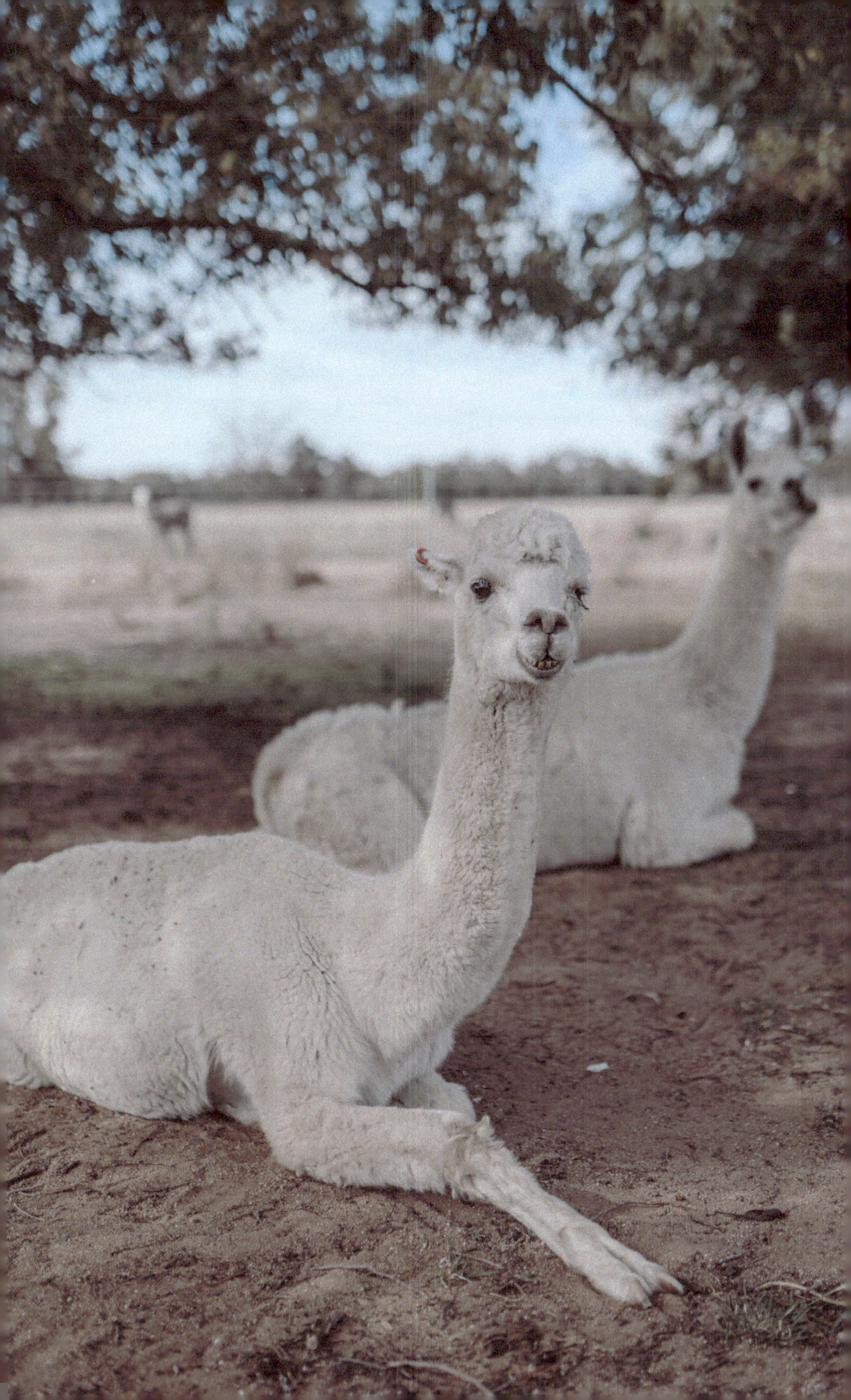

Afterword

Lessons from the Paddock

The alpacas teach me something new most days — usually at the exact moment I'm convinced I've finally got things under control. It really is adorable how wrong I continue to be.

They don't worry about yesterday or panic about tomorrow.

They don't spiral, catastrophise, or lie awake at 3 a.m. wondering whether they said something awkward three years ago.

They stay present. They trust their instincts. They trust the herd.

And they hum their way through the messy bits like it's the most natural thing in the world.

They remind me to slow down.

To look for the small good thing in any day.

To show up, even when I'd much rather hide.

To stop trying to wrestle the universe into obedience.

And occasionally, they remind me that spitting — while effective — is not an advisable communication tool for humans. Not for me anyway.

For the longest time, I thought I was here to take care of them.

Turns out, they've been taking care of me — teaching me patience, resilience, timing, humour, and the fine art of getting over myself — one completely unplanned lesson at a time.

If these pages have given you even a tiny piece of that — a laugh, a breath, a shift in perspective, or a moment that felt like a warm cuppa for your soul — I'm glad you're here. Truly.

And if, somewhere along the way, the alpacas start teaching you things too? Well... welcome to the herd.

About the Author

Amee Dennis is a farmer, fibre artist, and storyteller who somehow thought it was a brilliant idea to build a creative business around animals that spit. She spends her days juggling art, alpacas, and an ever-growing to-do list — mostly trying to look like she knows what she's doing while the alpacas quietly judge her.

From her home at Quentin Park Alpacas & Studio Gallery in rural South Australia, Amee works in the wonderfully messy space where creativity meets agriculture. She transforms the fleece grown by her own alpacas into handmade pieces that reflect the textures, colours, humour, and imperfections of the land she calls home.

Her writing leans into the honesty of farm life — the moments that never make the brochure but somehow make the story. Amee's work has been exhibited nationally and internationally, though she's still happiest on the farm with her daughter Emma, a cast of unpredictable animals, and a firm belief that coffee should always come before chores... and wine should always come after.

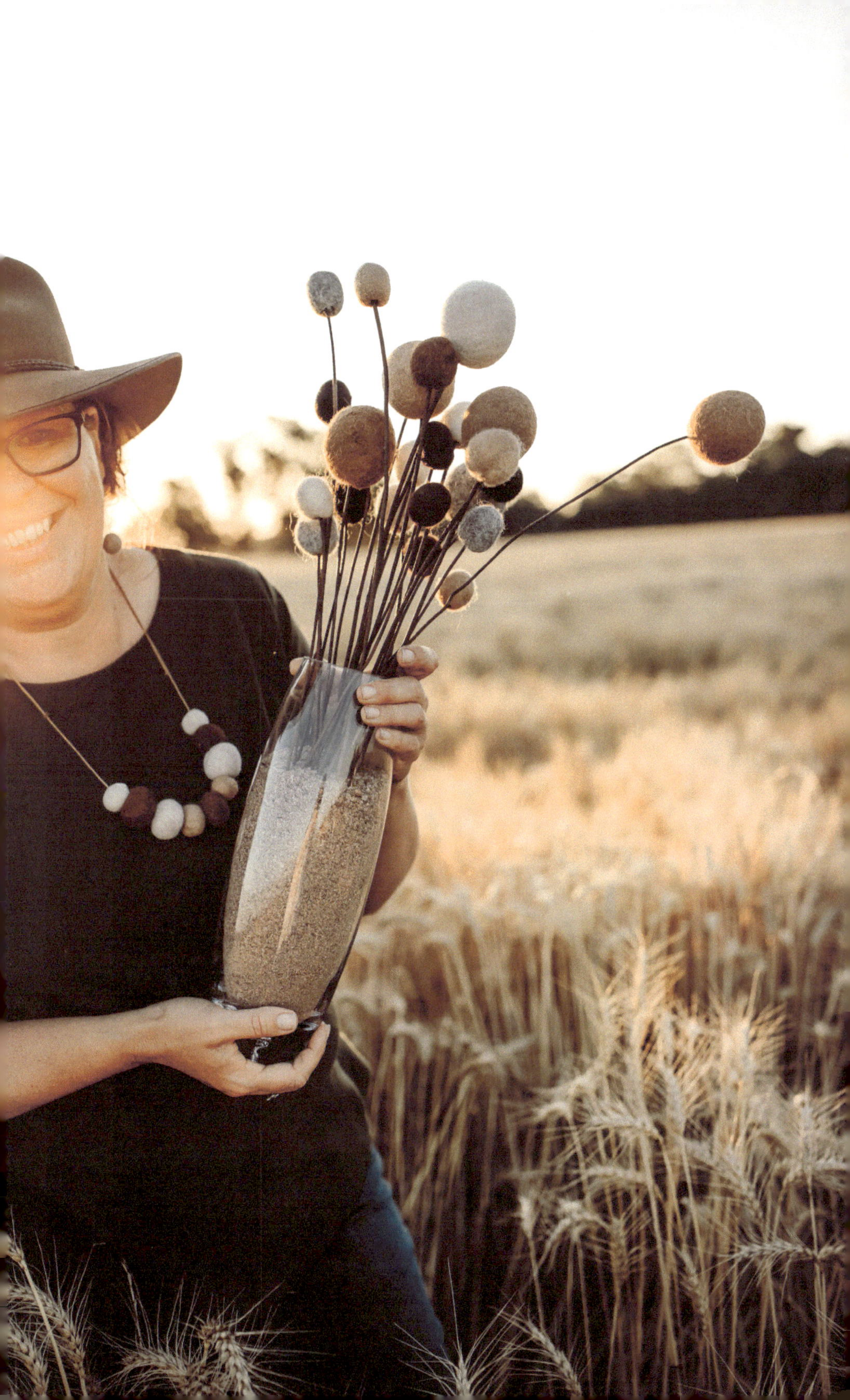

Now what?

Staying Connected

You're still here?
Thank you — truly — for becoming part of our herd.

I hope somewhere in these pages you found a moment you needed: a little smile, a breath, a shift in perspective, or at least a "good thing today."
If you'd like to keep wandering the paddocks with us (and honestly, why wouldn't you?), here are a few ways to stay connected:

Join the mailing list

It's where I share what's happening on the farm — upcoming events, product releases, behind-the-scenes chaos, and the things I probably shouldn't admit in public.
Don't worry, I'm terrible at spamming. Much to the marketing lady's despair, I usually remember to email every few weeks.

Visit the website

You'll find our paddock-to-product alpaca range, farm experiences, event updates, and all the things we're dreaming up next.

www.quentinparkalpacas.com

Come to the farm

Book an Alpaca Meet & Greet or grab a feed bowl.

Wander the seasonal gardens.

Explore the gallery and the treasures in the Farm Shop.

Stay in our off-grid farm stay or campsites.

Or simply sit with a cuppa from the café and enjoy the peace and the farm vista — before something inevitably escapes.

Keep reading

The Alpaca Perspectives Journal – A companion journal to help you explore your own notes from the paddock — the small good things, the shifts in perspective, the stories worth remembering.

Titty Tanning and Other Outrageous Antics – My next book, filled with the wild, ridiculous, heart-thumping, completely true stories of the alpacas and the farm.

Follow along

Instagram: **@quentinparkalpacas**

Facebook: **Quentin Park Alpacas & Studio Gallery**

TikTok: **@quentinparkalpacas**

YouTube: **@farmerandfibreartist**

However you choose to stay connected — online, in person, or simply by picking this book up again someday — We're grateful you're here.

Love

Amee and the 'Pacas

Acknowledgements

The people and animals who make it possible

There's no such thing as doing farm life — or writing a book — alone. (Trust me, I've tried. It didn't go well.)

To my family — thank you for loving me through the dust, the deadlines, the doubts, and the kind of chaos you can only get when animals are involved.

To my daughter, Emma — thank you for being the kind of human who shows up, who understands the heart of things, who can read alpaca body language better than anyone, and who keeps choosing this life with me. I'm endlessly proud of the incredible human you are.

To Sarah — the daughter I didn't give birth to, but who grew up alongside this farm and let it shape her just as much as it shaped me. Watching you find your place here, learn its lessons, and grow into yourself has been one of the unexpected gifts of this whole journey. Thank you for being part of our world — and part of my heart.

To the friends who have shown up with coffee, tools, wine, or a willingness to hold something odd while I "just fix this one thing" — thank you. Your patience and humour have saved me more times than I can count.

To the unexpected heroes who walked in at exactly the right moment — I see you. I'm grateful for you.

To every visitor who wandered in as a stranger and left as part of the herd — thank you for loving this place as much as we do.

To Miss Nicole — thank you for seeing the soul of this farm through your lens and capturing all the bits I couldn't put into words. You continue tell this story in a way only you could.

And lastly, to the 'pacas — thank you for the lessons, the laughter, the chaos, and the life worth writing down.

www.ingramcontent.com/pod-product-compliance
Lightning Source LLC
LaVergne TN
LVHW051134210926
848482LV00006B/317
* 9 7 8 1 7 6 4 4 3 4 8 0 5 *